A. W. Scott

Mammalia, Recent and Extinct

An Elementary Treatise for the Use of the Public Schools of New South Wales

A. W. Scott

Mammalia, Recent and Extinct
An Elementary Treatise for the Use of the Public Schools of New South Wales

ISBN/EAN: 9783337275945

Printed in Europe, USA, Canada, Australia, Japan

Cover: Foto ©berggeist007 / pixelio.de

More available books at **www.hansebooks.com**

MAMMALIA,

RECENT AND EXTINCT;

AN ELEMENTARY TREATISE FOR THE USE OF THE PUBLIC SCHOOLS OF NEW SOUTH WALES.

BY

A. W. SCOTT, M.A.

SYDNEY: THOMAS RICHARDS, GOVERNMENT PRINTER.

1873.

—

PINNATA.

SEALS, DUGONGS, WHALES,

&c., &c., &c.

PREFACE.

THE following pages, briefly descriptive of the economy of Seals, Dugongs, and Whales, and of their principal fossil allies, form the 2nd Part, or Section B, of an "Elementary Treatise on the Mammalia," designed for the use of the more advanced pupils in the Public Schools of this Country under the direction of the Council of Education.

In issuing this portion of the work in question before the 1st Part or Section A, I feel that I have travelled out ot the order of arrangement, as specified in the proposed synopsis of the Mammalian group in page 2 ; some remarks, therefore, by way of explanation for leaving the "unguiculata" to be hereafter dealt with, become necessary.

Whatever information we possess upon the natural history of the finned mammals, particularly in a popular, yet scientific form, has been so scantily and unequally distributed, that in this direction a comparatively new field may be said to be open to the teacher as well as to the youthful inquirer.

Influenced, also, by the great commercial value of several species of the pinnata, I have felt anxiously desirous to direct without further delay the attention, and thus possibly secure the sympathy, of readers, other than students, to the necessity of prompt legislative interference, in order to protect the oil and fur producing animals of our hemisphere, or at least some of them, against the wanton and unseasonable acts committed by unrestrained traders; and thus not only to prevent the inevitable extermination of this valuable group, but to utilize their eminently beneficial qualities into a methodical and profitable industry.

Keeping steadily in view these two objects, whose importance, I trust, will bear me out in deviating from my original intention in the order of the issue of publication, I have endeavoured—firstly; to interest the youthful mind with selections of well authenticated anecdotes of the general habits of these peculiar animals, accompanied, however, by those drier details of structural characters, essentially requisite to assist the more advanced and thoughtful student to a better understanding of the generic distinctions, and to aid him as a work of reference, or descriptive catalogue, should he be disposed in after-life to prosecute his researches in this difficult and imperfectly understood branch of Zoology, and—secondly; by devoting as much space as my limits would permit to the consideration of the animals whose products are of such commercial value to man, and whose extinction would so seriously affect his interests, to point out the pressing necessity that exists for devising the means of protection for the Fur Seals and the Sperm and Right Whales of the Southern Ocean.

To evidence what great results may be effected by considerate forethought, I refer the reader to pages 8 to 13 of this treatise, where he will see that, under the fostering care of the United States Government, the Northern Fur Seals of commerce, which but a few years ago were nearly extinct, have already, by their rapid increase and mild disposition, developed themselves into a permanent source of national wealth.

The islands of the Southern Seas, now lying barren and waste, are not only numerous, but admirably suited for the production and management of these valuable animals, and need only the simple regulations enforced by the American Legislature to resuscitate the present state of decay of a once remunerative trade, and to bring into full vigour another important export to the many we already possess.

I have cordially to thank the many friends, some residing at a considerable distance from Sydney, who, from time to time, and at much inconvenience, have either supplied me liberally with the loan of numerous scientific works for reference, or freely offered valuable information, acquired by personal experience, of the habits of some of the Southern animals of this section; as well as others to whom I am indebted for their considerate advice and material assistance during the progress of the compilation of this treatise.

In the present incomplete state, arising from the issue of one part only of the proposed publication, and which first portion, indeed, must be deemed as only on trial as a preliminary effort, I think it but right to withhold the names of the gentlemen alluded to, until the work in a more advanced stage has undergone the ordeal of public opinion—a course of procedure, I feel, greatly more to their advantage than to mine, whether that opinion be favourable or otherwise.

Sydney, 21 July, 1873.

TABLE OF CONTENTS.

KINGDOM—ANIMALIA.

* Organs arranged in pairs.

SUB-KINGDOM I. VERTEBRATA.

Animals which possess an internal skeleton, of which a back-bone, or vertebral column, composed of numerous joints, is always present. Blood red; sexes distinct; limbs never exceed two pairs.

† *Pulmonata, respiring by lungs.*

CLASS 1.—MAMMALIA, blood warm.—Man, Beasts, Dugongs, Whales, &c.

„ 2.—AVES, blood warm.—Birds.

„ 3.—REPTILIA, blood cold.—Lizards, Snakes, Tortoises, Frogs, &c.

†† *Branchiata, respiring by gills.*

„ 4.—PISCES, blood cold.—Fish.

SUB-KINGDOM II. ANNULOSA.

Animals without an internal skeleton: protected externally by a more or less hard articulated integument, and whose bodies and limbs are divided into segments or joints. Blood colourless, cold.

CLASS 1.—ARTICULATA.—Insects, Centipedes, Spiders, Crabs, &c.

„ 2.—ANNULATA.—Worms, Leeches, Intestinal Worms, &c.

SUB-KINGDOM III. MOLLUSCA.

Animals invertebrated, inarticulated, whose bodies are soft, and usually protected by a hard, calcified shell. Blood colourless, cold.

CLASS 1.—CEPHALOPODA.—Squid, Argonaut, Cuttle-fish, Nautilus, &c.

„ 2.—GASTEROPODA.—Periwinkle, Ear-shell, Land and Sea Snails, &c.

„ 3.—PTEROPODA.—Hyalæa, Clio, &c.

„ 4.—BRACHYOPODA.—Lamp-shells, Lingula, &c.

„ 5.—CONCHIFERA.—Oyster, Mussel, Cockle, &c.

„ 6.—TUNICATA.—Sea-squirts, Ascidians, Social Ascidians, &c.

B

** Organs not arranged in pairs.

Sub-kingdom IV. RADIATA.

Animals whose organs are placed in a circle around the mouth or axis of the body, and which consequently exhibit no right nor left side. Blood colourless, cold.

Class 1.—ECHINODERMATA. — Sea-cucumber, Trepang, Sea-urchin, Star-fish, &c.

„ 2.—ZOOPHYTA.— { Sea-nettle, Portuguese Man-of-war, Coral and Sponge Animals, countless aquatic microscopic beings, &c.

Sub-kingdom I. VERTEBRATA.

Class 1.—MAMMALIA,
or Animals which suckle their young.

Sub-class I. PLACENTALIA.

Animals provided with a placenta, by which the fœtus is connected to the womb and nourished through the medium of the blood.

Section A. *UNGUICULATA.*

Limbs furnished with nails or claws.

Order 1.—BIMANA.—Man.

„ 2.—QUADRUMANA.—Old and New World Apes, Marmozets, Lemurs, &c.

„ 3.—CHEIROPTERA.—Bats, Flying Fox.

„ 4.—INSECTIVORA.—Hedgehog, Shrew, Mole, &c.

„ 5.—RODENTIA.—Rat, Porcupine, Hamster, Squirrel, Rabbit, Beaver, &c.

„ 6.—CARNIVORA.—Bears, Dog, Cat, Weasel, Otter, Sea Otter, &c.

Section B. *PINNATA.*

Limbs fin-shaped.

ORDER 7.—PINNIPEDIA.—Eared and common Seals, Walrus, Sea Elephant, &c.

„ 8.—DEINOTHERIA.—Deinotherium, Toxodon.

„ 9.—SIRENIA.—Manatee, Dugong, &c.

„ 10.—ZEUGLODONTIA.—Zeuglodon, &c.

„ 11.—CETACEA.—Whales, Dolphins, Porpoises.

Section C. *UNGULATA.*

Limbs furnished with hoofs.

„ 12.—ARTIODACTYLA.—Hippopotamus, Pig, Camel, Deer, Sheep, Ox, &c.

„ 13.—PERISSODACTYLA.—Horse, Tapir, Rhinoceros, Hyrax, &c.

„ 14.—PROBOSCIDEA.—Elephant, Mastodon, &c.

Section D. *SUB-UNGULATA.*

Limbs furnished with hoof-like claws.

„ 15.—EDENTATA.—Megatherium, Sloth, Aard-vark, Armadillo, Ant-eater, &c.

SUB-CLASS II. IMPLACENTALIA.

Animals destitute of the placenta; the fœtus not connected to the womb.

ORDER 16.—MONOTREMATA.—Australian Hedge-hog, Duck-billed Platypus.

„ 17.—MARSUPIALIA.—Kangaroo, Wombat, Opossum, Native Cat, &c.

Section B. *PINNATA.*

Seals, Dugongs, and Whales, the living representatives of the Pinnata, although they differ greatly in many essentials of structural character and external form, appear nevertheless to constitute a very compact sectional group, being intimately allied—not only by the several usually recorded features of similarity, progressively exhibited by a series of transitional links,—but likewise, and that impressively, by the possession in common of a peculiarity of organisation, unknown to any other of the Mammalian Orders, with a solitary exception.

The following comparative details will serve to illustrate the former assertion, but the restricted nature of this treatise permits me no license in this, as well as in other cases, to diverge into anatomical particulars, beyond the simple allusion to the existence of the latter; it being imperative to refer to the fact, in order to sustain, by its additional contribution, the validity of the present deviation from the beaten track, in allying, so immediately, the Pinnipedia with the Sirenia and Cetacea.

The transition, apparently so incongruous, from the purely terrestrial quadrupeds to such mammals as the Dugong and the Whale, whose hinder members are defective, and whose lives are rigorously restricted to the waters, is effected through the medium of the semi-aquatic carnivorous animals, the Sea Otter, the last erratic representative of the land flesh-eaters, and the Sea Bears, the first of the series of the present section.

The lengthened fur-covered body, the shortened limbs, still capable of being used for progression on land,—the palmated and unguiculated feet,—the small, somewhat depressed tail,—the oceanic habits,—the similarity of food, and mode of capture,—and the process of mastication commonly above the surface of the water,—when considered in the aggregate, attest the affinity of the Sea Bear to the Sea Otter in sufficiently marked characters, however greatly each separate function may be modified under the varying conditions of their existence.

Then, by the more elongated form of the Phocidæ propriæ,—by the nearly immovable condition of their hinder limbs, which stretch rigidly backwards almost in a line with the body, the broad webbed feet being capable only of lateral free motion for the purpose of propulsion through the water,—and by the external position of the nostrils at the end of the muzzle,—the animals of the order Pinnipedia approach those of Sirenia.

The Manatee, Dugong, and the Rhytina, in possessing horizontal, cartilaginous, tail-like, hinder extremities,—in having the nasal apertures placed high up on the skull, although the nostrils terminate at the extremity of the face,—in many other portions of their structural character,—and in their wholly aquatic existence,—graduate so naturally into the whale-tribe, as to have caused Cuvier to term them the herbivorous Cetacea.

These inferences, derived comparatively from only a few generalities, may be considered as too theoretical, but palæontologists have already revealed to us the interesting forms—some unfortunately imperfect—of the extinct Deinotherium, the Halitherium, and the Zeuglodon, whose remains have, by their intermediate character, materially aided to substantiate the alliance between the finned Mammals.

There still exists, however, a wide and indefinite interval which separates the three orders, and which will have to be filled up by the further discoveries of fossil relics before any satisfactory linear arrangement can be assured.

PINNIPEDIA.

FAMILIES.	GENERA.	MOLAR TEETH.	EAR CONCHES.	UNDER FUR.	HIND LIMBS IN REPOSE.
Otariadæ......	Arctocephalus.. Zalophus......... Otaria	Single-rooted.	} External.	Abundant, long. Sparse. None.	} Bent forwards.
Trichechidæ.....	Trichechus ...... Macrorhinus ...			} None.	
Cystophoridæ {	Cystophora.......			} Moderate, short.	
	Monachus.........	More or less double-rooted.	} Not visible.		} Bent backwards.
Phocidæ	Phoca .... Halichoerus..... Stenorhynchus.			} None.	

Order 7.   PINNIPEDIA,[1]

Amphibia[2] of Cuvier, Otaries,[3] Walrus,[4] Seals.[5]

The front limbs of all the animals which compose this order are powerful, short, nearly hidden within the skin of the body ; the paws, however, advance, are fin-like, and provided with five long fingers, which diverge from each other, and are completely embedded in the surrounding membrane: these fingers, in general, diminish in size from what we may call the thumb to the little finger.

The fore-limbs are used for swimming purposes, for seizing the prey, for assisting in movements on land and for ascending rocks or blocks of ice.

The hind limbs are even more powerful than the front ones, and when at rest are in some species directed forwards, similar in position to those of terrestrial mammals; in others backwards, in a line with

---

[1] *Pinna*, a fin ; and *pes*, a foot.

[2] Ἀμφίβιος (ἀμφὶ and βιός), capable of living on land and in water. I may remark that the term amphibious, when applied to these animals, is incorrect, for not possessing gills they cannot breathe under water, but must come to the surface to respire the atmospheric air, as other mammals do. Existing, however, on fish and other marine prey, they possess, on extraordinary occasions, the essential attribute of only breathing once in twenty minutes, whereas many land animals are compelled to do so twenty times in the minute.

[3] Οὖς, ὤrbs, the ear.

[4] From the German "*wall*," as in *wallfisch*, a whale, and "*ross*" a horse.

[5] From the Saxon "*seel*," "*sele*," "*syle*."

the body, which they terminate; the bones are short, and strong; the five toes of the foot are filled up between with a flexible membrane, which enables them to spread out when in action into broad webbed paddles, and again in repose to fold together; of these toes the lateral ones are the largest, the others diminishing towards the centre.

By means of these hind limbs, seals are principally rendered expert swimmers, and perform their evolutions in the water with ease, rapidity, and endurance.

The body is elongated, conical, and tapers from the chest to the tail; it is clothed either with long, soft, compact hair, enveloping a valuable under-fur,—or with hair short, smooth, firmly adpressed to the skin, and slightly unctuous. The mammæ are ventral. The head is round, with a large, full, fleshy muzzle, studded with long stiff bristles. The eyes are large and dark, expressive of intelligence, and eminently adapted for seeing under water. The ears are very small, mostly not visible externally. The neck is long and flexible; the cervical vertebræ, free. The sternum is usually composed of eight bones, to which nine or ten pair of ribs are directly joined. The costo-sternal ribs are cartilaginous. The dorsal line is without any protuberance. The tail is very short, usually compressed, and placed immediately between the hind feet.

As might be expected from this peculiar structure, so admirably adapted for the watery element in which they pass a great portion of their lives, these animals when on the land are very ungainly in their movements. It is only in a few species where progression appears to be accomplished, though very awkwardly, in a manner similar to that of the terrestrial quadruped; while in others, it is attained by bending or arching the extremely flexible back-bone, by fixing firmly the posterior portion of the body on the ground, and then by suddenly straightening out, in front, the whole frame. By a quick repetition of this movement, a series of jerking leaps takes place, and, assisted materially by the fore-paws, a speed is attained, especially on the ice, sufficient to outstrip a man running in pursuit.

Seals are eminently gregarious, and consequently are seldom met with except in large herds. They resort to the land for the purpose of bringing forth and suckling their young—which at a birth is commonly one, very rarely two—for basking in the sun, in the warmth of which they delight, for repose and slumber during the night, and for shelter from tempestuous weather.

To ascend rocks or masses of ice of ordinary elevations, they fasten their fore-paws, with the gripe of a vice, on inequalities, and uplifting their unwieldy carcasses, they with tolerable facility gain the summit; but when the sides of these elevations prove too precipitous, they await the swell of the wave, which wholly or partially floats them to their purposed place of repose; but in the latter case they cling with tenacity to the face of the rock until another and larger wave lifts them to a sufficient height.

The brain of the Seal tribe is usually much developed, and writers best acquainted with the habits of the species accord to these animals the possession of a considerable amount of intelligence and sagacity, scarcely inferior to those exhibited by the dog. This favourable opinion has been frequently verified by many interesting examples, while in a state of semi-domestication; although it is palpable that these faculties, when exercised in their natural element, the full extent of which we can have no means of accurately ascertaining, must necessarily excel those which they manifest on shore.

With the view of freeing from complication the many intricacies which at the present time obscure the consideration of the animals composing this imperfectly understood Order, and to insure to the student a ready, yet comprehensive insight into the systematic disposition of the species, I commence by rejecting, as comparatively valueless, the highly elaborated Synopsis of Tribes, Genera, and Species, which have been solely based upon the slight variations exhibited in the cranial development; for such indications in the main are unreliable, and their omission, in regard to methodical determination, presents no appreciable obstruction in the way of future research.

By the material curtailment, which this decision facilitates, of the list of those alleged distinct kinds whose identity rests wholly upon such adventitious qualities, and likewise by uniting under the same genera the animals whose separate positions have been established upon the equally trivial evidence of a shade of colour or of a limited range of habitat, I arrive at a simpler, and, I believe, at a truer estimate of the number of species which constitute the Pinnipedian group.

The Seals are arranged in this elementary treatise under two heads,—the Eared, and the Earless Seals: the former represented by one family, the Otariadæ; and the latter by three, the Trichechidæ, Cystophoridæ, and Phocidæ.

SEALS *with* external Ears.

Family I. OTARIADÆ.[1]

Incisors $\frac{3\cdot3}{2\cdot2}$, canines $\frac{1\cdot1}{1\cdot1}$, molars $\frac{5\cdot5}{5\cdot5}$, or $\frac{6\cdot6}{6\cdot5} = 34$ or 36.[2]

The four middle upper incisors frequently have double cutting edges;[3] the lower ones are bifurcate; molars, generally closely approximated, are

---

[1] οὖς, ὠτὸς, the ear.

[2] This dental formula is the usual concise mode of describing the number and position of the various teeth. The upper figures refer to the teeth of the upper jaw, and the lower ones to those of the under jaw; while the hyphen serves to distinguish the right from the left side.

[3] "A circumstance hitherto unknown in any other animal."—CUVIER.

conical, with very large single roots; in some the last of the upper ones have two roots, and small, compressed, lobed crowns; head short, dog-like; muzzle enlarged, and furnished with strong, stiff whiskers; ears provided with a sub-cylindrical external conch ; eyes large, protected by eye-lids; mouth very large; tongue forked at the extremity; fore limbs fin-like, situated far back; hind limbs rather produced, comparatively free, and bent forwards in repose,—the limbs evincing, by their freer use, a nearer approach to the terrestrial Carnivora than those of any other of the Seal tribe ; nails flat, small, slender; membrane of the feet prolonged beyond the nails into as many lobes as there are toes; tail short, conical; mammæ four, ventral; males much larger and darker in colour than the females.

These animals, during progression on land, walk on their fore and hind limbs, and in repose turn the hind feet forwards. In habits they are gregarious and polygamous.

SEALS, *Adults*, with abundant under-fur.

Genus ARCTOCEPHALUS,[1] F. Cuvier.

Incisors $\frac{3-3}{2-2}$, canine $\frac{1-1}{1-1}$, molars $\frac{6-6}{5-5} = 36$.

Upper incisors large, lower ones small; canines large, sharply pointed. Head and face somewhat elongated; cerebral region slightly elevated; sagittal crest moderately developed; muzzle narrow, pointed, moderately enlarged between and above the nostrils ; body more slender, feet and toe-flaps proportionately longer, than those of the Sea Lion; claws very small, scarcely visible; toes of the hind feet short, all nearly of the same length; body covered with hair, and with thick, permanent, under-fur. In size the Sea Bears are much smaller than the Sea Lions.

ARCTOCEPHALUS URSINUS,[2] Linnæus.   Northern Fur-Seal of Commerce.

Synonyms—*Phoca ursina*, Linn.
    Otaria ursina, Péron. Desm.; Nilsson, Gray, Peters.
    Arctocephalus ursinus, Gray, Gill.
    Arctocephalus Californianus, Gray, B. M. C. 1866, p. 51.
    Callorhinus ursinus, Gray, B. M. C. p. 44; Suppl. 1871, p. 14 ; Allen, Bull. Mus. Comp., vol. ii., p. 73.

The Northern and Southern Fur-Seals are considered by Dr. Gray to be generically distinct; the skull of the former (Callorhinus)[3] "being easily known" from the latter (Arctocephalus) " by the shortness of the face, and convexity of the nose."

It must be borne in mind, that even in the same species the development of the skull exhibits marked sexual characters, as well as many of those differences of form which occur during the various periods of growth. So frequently are these cranial variations met with, that it

---

[1] ἄρκτος, a bear, and κεφαλή, the head—bear-headed.
[2] *Ursinus*, bear-like.
[3] καλλος beautiful, and ῥινος skin.

becomes almost impossible to nicely discern the relative position of individuals, even under favourable circumstances, and the difficulty is greatly enhanced by the imperfect data afforded by the examination of a few specimens only.

The exceptional form of the cranium of the Northern Fur-Seal, as quoted above, appears to display no characters more strongly defined than those commonly seen in the skulls of many species of the same genus among the undomesticated mammalia; consequently, a specific distinction applied to this animal would probably have been quite sufficient to meet every requirement for scientific classification.

To the foregoing doubts as to the propriety of generic separation from the Southern Fur-Seal, I may further add, that little dependence, for the purpose of distinguishing kinds, can be placed on the appearance of the skin, or on the limit of the range of habitat; for the colouring of the external hair, and the length, abundance, and quality of the under-fur, are greatly diversified by sex, by age, by seasonal condition, and by climate. And the geographic range is not confined within small bounds, but on the contrary it is extensive, as clearly established by the habitat of this Fur-Seal, which extends from the shores of Kamtschatka to those of California,[1] an extent of ocean greater than that from California to the Island of Juan Fernandez, or than those intervening spaces between the numerous localities in the Southern Seas, the recognised strongholds of the Southern Fur-Seals. The barrier, therefore, if *any*, which forbids the intercourse between these antipodean relatives exists, not in the distance, but in the passage across the warm temperature of the torrid zone.

Taking such a view, I can scarcely accede to so broad a separation as that proposed by Dr. Gray, but I am willing to consider the Northern animal as a distinct species,—suggesting, however, the probability of its ultimately proving to be only a member of the one great Fur-Seal tribe of both hemispheres.

The colour of the external coating of the male, when adult, varies from black-grey to brown-grey, occasionally pure black; while the adult female is usually grey, or ash-coloured, but during the shedding of the coat, many are seen partly ash-coloured and partly brown. The young of both sexes, previous to the first moult, are uniformly glossy black, silvered more or less by short white tips; mostly so about the nape of the neck and hinder parts of the body; or, as Dr. Gray himself observes, "the skin is so like that of the Arctocephalus nigrescens,[2] that we were induced to regard it as a second specimen of that species."

The under-fur of both sexes and of the young is of a rich reddish colour, more or less tinted with deeper or lighter shades.

---

[1] The Northern Sea Lion (Otaria Stelleri) and the Zalophus Gillespii, also occupy precisely the same extensive range.
[2] Arctocephalus Falklandicus of *Peters, Allen, Sclater,* &c.

The males, when aged, will reach to 8 feet in length, but animals of 6 feet, or slightly under, are most frequently met with. The females are very much smaller, scarcely ever exceeding 4 feet.

It will be seen, when I treat of the Southern Fur-Seal, that this description of the size, the colour of the hair, and under-fur of the Northern animal is applicable to both, and, in an account of the habits of the present species, those of the Antarctic Fur-Seal will be found to be equally truthfully depicted.

"This creature," writes Steller, "has four feet on which it can walk and stand, somewhat like land animals ;" " when on shore, with the hind feet folded under, it plants the paws in front and sits as dogs often do, so that the toes perform the office of heels." "These animals are found in amazing numbers in the Islands of the North-west Coast of America, and so crowd the shore that they oblige the traveller to quit it, and scale the neighbouring rocks." They are as regularly migratory as birds of passage." "They live in families, every male being surrounded by from eight to fifty females, which he guards with the jealousy of an Eastern monarch. Each family keeps separate from the others, notwithstanding they lie in thousands along the shore,—every family, including the young, amounting to about 100 or 120. Even at sea the distinctness of the families may be perceived." "When fighting they utter hideous growls,—when amusing themselves they low like a cow,—and after victory chirp like a cricket,—and upon receiving a wound complain like a whelp."

"Some twenty or thirty years ago there was a most wasteful destruction of the Fur-Seal, when young and old, male and female, were indiscriminately knocked on the head. This improvidence, as every one might have expected, proved detrimental in two ways. The race was almost extirpated ; and the market was glutted to such a degree, at the rate, for some time, of 200,000 skins a year, that the prices did not even pay the expenses of carriage. The Russians, however, have now adopted nearly the same plan which the Hudson's Bay Company pursues in recruiting any of its exhausted districts, killing only a limited number of such males as have attained their full growth, a plan peculiarly applicable to the Fur-Seal, inasmuch as its habits render the system of husbanding the stock as easy and certain as destroying it. In the month of May, with something of the regularity of the almanac, the Fur-Seals make their appearance at the Island of St. Paul, one of the Aleutian Group. Each old male brings a herd of females under his protection, varying in number according to his size and strength ; the weaker brethren are obliged to content themselves with half-a-dozen wives, while some of the sturdier and fiercer fellows preside over harems that are 200 strong. From the date of their arrival in May, to that of their departure in October, the whole of them are principally ashore on the beach. The females go down to the sea once or twice a day,—while the male, morning, noon, and night, watches his charge with the utmost jealousy, postponing even the pleasures of eating and drinking, and sleeping, to the duty of keeping his favourites together. If any young gallant ventures by stealth to approach any senior chief's bevy of beauty, he generally atones for his imprudence with his life, being torn to pieces by the old fellow,—and such of the fair ones as may have given the intruders any encouragement are pretty sure to catch it in the shape of some secondary punishment." "At last the whole herd departs, no one knows whither."

"The mode of capture is this:—At the proper time, the whole are driven like a flock of sheep to the establishment, which is about a mile distant from the sea; and there the males of four years, with the exception of a few that are left to keep up the breed, are separated from the rest and killed. In the days of promiscuous massacre, such of the mothers as have lost their pups would ever and anon return to the establishment, absolutely harrowing up the sympathies of the wives and daughters of the hunters, accustomed as they were to the scene, with their doleful lamentations."—*Sir George Simpson.*[1]

"The male Fur-Seal does not attain mature size until about the sixth year. He then measures in total length from seven to eight feet, and six and seven in girth. His colour is then dark brown, with grey overhair on the neck and shoulders. When in full flesh his weight varies from five to seven hundred pounds. These and no others occupy the rookeries (or breeding grounds) with the females.

"A full-grown female measures four feet in length and two and a half around the body. She usually weighs from eighty to a hundred pounds. Her colour, when she first leaves the water, is a dark steel-mixed on the back, the sides and breast being white; but she gradually changes somewhat, and in eight or ten days after landing becomes dark brown on the back, and bright orange on the breast, sides, and throat.[2] Hence, it is easy to distinguish those that have just arrived from those that have been several days on shore. The female breeds the third year, and is full-grown at four years.

"The breeding rookeries,[3] which are frequented exclusively by the old males and females with their pups, occupy the belt of loose rocks along the shores between the high-water line and the base of the cliffs or uplands, and varies in width from five to forty rods. The sand beaches are used only as temporary resting-places, and for playgrounds by the younger seals; these beaches being neutral ground, where the old and infirm or the wounded may lie undisturbed.

"The old male appears to return each year to the same rock, so long as he is able to maintain his position. The native chiefs affirm that one seal, known by his having lost one of his flippers, came seventeen successive years to the same rock.

"Those under six years are never allowed by the old ones on these places. They usually swim in the water along shore all day, and at night go on the upland above the rookeries and spread themselves out, like flocks of sheep, to rest.

---

[1] "Narrative of a Journey round the World in 1841 and 1842."
[2] See description of female of the Southern Fur-Seal by Musgrave and Morris, page 15.
[3] Pribyloff group of Islands.

" Wherever a long continuous shore-line is occupied as a breeding rookery, neutral passages are set apart at convenient distances, through which the younger seals may pass from the water to the upland and return unmolested. Often a continuous line moving in single file may be seen for hours together going from the water to the upland, or the reverse, as the case may be. When suddenly disturbed while sleeping on the upland by an attempt of an animal to cross the rookery at any other place, a general engagement ensues, which often results in the death or serious crippling of the combatants.

"The old males are denominated by the natives *Seacutch* (married seals). These welcome the females on their arrival, and watch over and protect them and their young until the latter are large enough to be left to the care of their mothers and the younger males.

" Those under six years old are not able to maintain a place on the rookery, or to keep a harem, and these are denominated *Holluschuck* (bachelors).

" As soon as a female reaches the shore, the nearest male goes down to meet her, meanwhile making a noise like the clucking of a hen to her chickens. He bows to her and coaxes her until he gets between her and the water, so that she cannot escape him. Then his manner changes, and with a harsh growl he drives her to a place in his harem.

"Then the males higher up select the time when their more fortunate neighbours are off their guard to steal their wives. This they do by taking them in their mouths and lifting them over the heads of the other females, and carefully placing them in their own harem, carrying them as cats do their kittens.

" Frequently a struggle ensues between two males for possession of the same female, and both seizing her at once pull her in two, or terribly lacerate her with their teeth.

" In two or three days after landing, the females give birth to one pup each, weighing about six pounds. It is entirely black, and remains of this colour the whole season.

" There are at least twelve miles of shore line on the Island of St. Paul's[1] occupied by the seals as breeding grounds, with an average width of fifteen rods. There being about twenty seals to the square rod, gives 1,152,000 as the whole number of breeding males and females ; deducting one-tenth for males, leaves 1,037,800 breeding females. Allowing one-half of the present year's pups to be females, this will add half a million of breeding females to the rookeries of 1872, in addition to those now there, while the young of last year and the year before are also to be added. This estimate does not include the males under six years of age, those not being allowed on the rookeries by the older males, nor the yearlings. If we now add those frequenting St. George's Island, which number half as many, and make a very liberal discount for those that may be destroyed before reaching maturity, the number is still enormous. It will also be seen that the

[1] Coast of Kamtschatka.

great importance of the Seal fishery is not to be calculated from the basis of its present yield, since each year adds to its extent, as with proper care the number can be increased until both islands are fully occupied by these valuable animals.

" Previous to 1866 these skins were worth only three dollars each, but owing to recent improvements in their manufacture they have become fashionable for ladies' wear, and soon after the transfer of the Territories to the United States the price rose to seven dollars." [1]

ARCTOCEPHALUS FALKLANDICUS, Shaw. The Southern Fur-Seal of Commerce.

Synonyms—*Falkland Island Seal*, Pennant.
Phoca Falklandica, Shaw.
Phoca antarctica, Thunberg.
Otaria Falklandica, Desmarest.
Otaria cinerea, Péron. Desm.; Peters.
Otaria Delalandii, F. Cuvier.
Otaria Falklandica, ♀ or young. Jardine's "Nat. Libr."
Arctocephalus antarcticus, Gray, S. and W. and Suppl. Cape Fur-Seal.
Arctocephalus cinereus, Gray. Australian Fur-Seal.
Arctocephalus nigrescens, Gray. Southern Fur-Seal.
Arctophoca Philippii, Peters. Chilian Fur-Seal.

The few remarks in regard to the variations arising from the asymmetry of the cranial structure, and to the differences exhibited by external colouring of the skin, all referable to natural or incidental causes, which I presented for the consideration of the reader, when discussing the natural position of the Northern Fur-Seal, apply with equal force to the Fur-Seals which inhabit the Southern Seas; and although several of these animals are considered as distinct species by most writers, yet I have ventured to include them under the one kind, the Falkland Island Seal of Pennant and Shaw.

In adopting this course, it becomes necessary that some valid reasons should be given for departing so materially from the usual arrangement.

In instituting the following comparisons, I refer the student to the list of the alleged distinct species enumerated in the foregoing synonyms, and which are described by Dr. Gray, in his Catalogue of the Seals in the British Museum of 1866, and Supplement 1871; for these publications are of great authority, inexpensive, and more easily attainable here than any others with which I am acquainted.

First, then, as to the scientific value of those distinctions said to be so readily seen in the form of the skulls and in the minor differential points in the dentition, so as to constitute well defined species.

---

[1] On the Eared Seals (Otariadæ), by J. A. Allen and Charles Bryant. Aug., 1870.

I arrive, by a careful analysis of the researches of certain modern authors of great experience in such matters, at the following general but singularly conclusive results, namely :—

Dr. Peters[1] considers *nigrescens* and *Philippii* = *Falklandicus* of Shaw also *cinereus* = *antarcticus*.

Mr. Allen[2]	„	*nigrescens, cinereus, antarcticus, Forsteri* and *Philippii* =	do.
Dr. Burmeister[3]	„	*Philippii* =	do.
Mr. Sclater[4]	„	*nigrescens* =	do.
Capt. Abbott[5]	„	*nigrescens* =	do.

From these deliberate expressions of opinion, I am led to conclude that, if these zoologists are correct in their views, the whole series of the species of the Southern Fur Seals, defined with such precision in the British Museum Catalogue and Supplement, with the exception of Dr. Gray's Falkland Island Fur Seal, merge into the one, the Arctocephalus Falklandicus of Shaw ; if otherwise, the difficulty, even among the most proficient, of discriminating species correctly, is so palpably displayed, that not only a disagreeable impression of unreliableness for the method of determination is stamped on the mind, but that no positive conclusions can possibly be drawn from principles so imperfect in themselves, from the slight and inconstant nature of their characters.

Next, let the size and external colouring of the most familiarly known animals, which locate the numerous and by no means widely apart spots which stud the Southern Seas, be contemplated seriatim, and I think it will be admitted that the similarity, *inter se*, revealed by both of these properties, will likewise corroborate by their concurrent testimony the probable unity of the many so-called species.

The Sea Bears, which inhabit the *Falkland Islands*, "have the hair short, cinereous, tipped with dirty white ; length, 4 feet "—*Phoca falklandica, Shaw*. "Blackish-brown, grey-black"—*Arctoc. nigrescens, Gray*. "The full-grown seal is about the size of the common English seal ; the hair differs in colour, being sometimes grey, and at other times of a brownish tint ; that of the young is of a darker brown colour."—*Abbott*.

South Shetland Islands. "Nothing is more astonishing than the disproportion in the size of the male and female ; a large grown male is six feet nine inches, while the female is not more than three and a half feet ; the young are at first black, but after a few weeks they become grey."—*Weddell*.

Island of Juan Fernandez. "They are the size of an ordinary calf ; their hair is of different colours, as black, brownish-grey, and spotted. —*Dampier*. "Above dark-grey, more greyish on the head and neck, brownish-white beneath."—*Otaria Philippii.—Peters*.

[1] Professor W. Peters, Berlin.
[2] Professor J. A. Allen, Cambridge, America.
[3] Museum, Buenos Ayres.
[4] Secretary to Zoological Society of London.
[5] Proceedings Zool. Soc., Lond., 1868.

New Year's Island, Staten Land. "Their size was equal to that assigned them by Steller; their hair is dark brown, sprinkled with grey."—*Forster.* "They are rather larger than the common seal, and their general colour is iron-grey."—*Cook.*

Coast of Australia. "Black, greyer beneath; the hair changes its colour as the animal grows; the young being generally black; and the adult males and females also differ considerably in the colour of the hair."—*Macgillivray, Arct. cinereus.—Gray.*

Auckland Isles. "Males, bulls, or sea lions, uniformly blackish-grey; length usually six feet, but the aged animal greatly exceeds this size. Females, cows, or tiger seals, grey, golden-buff,' or beautiful silver colour, sometimes spotted like a leopard; smaller than the males."—*Musgrave.*

Auckland Islands, South Coast of New Zealand, Shetland, Antipodes, and *Chatham Islands.* "Adult male, or wig, uniformly blackish; pups born black; after a few weeks they become grey; at a year old the grey changes to light-brown, and when adult, to black, or blackish-grey. Adult female, or clapmatch, grey to silver-grey, at times golden-buff;[1] pups black."—*Morris.*[2]

South Africa. "Adult male, or large wig, hair whitish, intermixed with a few black ones; adult, or middling, hairs reddish-white, grizzled, with scattered black hairs; young, or black pup, black, without any grey tips. *Arct. antarcticus, from skins.*"—*Gray.*

From such perplexing sources no reasonable data for distinguishing species can be deduced. I would, therefore, advise the student to consider all the animals mentioned in the synonyms as of the one kind, the A. Falklandicus of Shaw; at least, until stronger proofs of dissimilarity be produced to displace the present characteristics, which certainly appeal more to the imagination than to reality.

The Arctocephalus Falklandicus may be thus described:—

The males, when aged, are whitish-grey, and between seven and eight feet in length; when adult, brown-grey to black-grey, and about six feet in length; young, grey, upper portions soon assume darker colours; pups, black.

The females, when adult, are ash-grey to silver-grey, at times golden-buff, frequently spotted; from three and a half to four and a half feet in length, even more when aged; pups, black.

The under-fur of both sexes is rich reddish, diversified by deeper or lighter shades, and variable in length and abundance; the whole being influenced by age, sex, and condition.

Habitat,—Southern Seas generally.

The economy of the Southern Fur-Seal has from time to time been luminously portrayed by many writers, and, as mentioned before in page 10, will be found, by perusal of the following extracts, selected from many, to correspond precisely with the habits entertained by the Northern animal.

[1] See Mr. Allen's description of the female of the Northern Fur-Seal, p. 11.
[2] For many years a sealer by profession, and now residing in Sydney.

By Mr. Forster, the companion of the celebrated Captain Cook, we are told, that at Staten's Land, where these animals existed in thousands, "as soon as I was near enough I shot the surly creature dead, and at that instant the whole herd hurried to the sea, and many of them hobbled along with such precipitation as to leap down between forty and fifty perpendicular feet upon the pointed rocks on shore without receiving any hurt, which may be attributed to their fat easily giving way, and their hide being remarkably thick." "The young cubs barked at us, and ran at our heels when we passed, trying to bite our legs."

Mr. Weddell informs us "When these Shetland Seals were first visited, they had no apprehension of danger from meeting men; in fact they would lie still while their neighbours were killed and skinned, but latterly they had acquired habits for counteracting danger by placing themselves on rocks from which they could in a moment precipitate themselves into the water." "Their sense of smell and hearing is acute, and in instinct they are little inferior to the dog." "These," the females, "in the early part of December begin to land, and they are no sooner out of the water than they are taken possession of by the males, who have many serious battles with each other in procuring their respective seraglios, and by a peculiar instinct they carefully protect the females under their charge during the whole period of their gestation. By the end of December all the females have accomplished the purpose of their landing." "By the middle of February the young are able to take the water, and after being taught to swim by the mother, they are abandoned on the shore, where they remain till their coats of fur and hair are completed." (From Jardine's "Nat. Libr.")

A detailed account of the habits of the Fur-Seal of the Auckland Isles has recently been given by Mr. Musgrave,[1] which he acquired during a compulsory residence in their midst of nearly twenty months. Of the females, he relates that "Their nose resembles that of a dog, but is somewhat broader; their scent appears to be very acute. The eyes are large, of a green colour, watery and lustreless; when on shore they appear to be constantly weeping." "In the latter part of December, and during the whole of January, they are on shore a great deal, and go wandering separately through the bush, and into the long grass on the sides of the mountains above the bush, constantly bellowing out in the most dismal manner. They are undoubtedly looking out for a place suitable for calving in. I have known them to go to a distance of more than a mile from the water for this purpose." "Females begin to breed when two years old, and carry their calves eleven months, and suckle them for about three months." "Before they have their calves, the cows lie sometimes in small mobs (from twelve to twenty), as well as while giving suck, and there is generally one or two bulls in each mob. The cows are evidently by far the most numerous." Of the habits of the very young, he says: "It might be supposed that these animals, even when young, would readily

---

[1] Narrative of the wreck of the "Grafton." Melbourne, 1865.

go into the water—that being one of their natural instincts—but strange to say such is not the case; it is only with the greatest difficulty and a wonderful display of patience that the mother succeeds in getting her young in for the first time. I have known a cow to be three days getting her calf down half a mile, and into the water; and what is most surprising of all, it cannot swim when it is in the water; this is the most amusing fact: the mother gets it on her back, and swims along very gently on the top of the water; but the poor little thing is bleating all the while, and continually falling from its slippery position, when it will splutter about in the water precisely like a little boy who gets beyond his depth and cannot swim. Then the mother gets beneath it and it again gets on her back. Thus they go on, the mother frequently giving an angry bellow, the young one constantly bleating and crying, frequently falling off, spluttering and getting on again; very often getting a slap from the flipper of the mother, and sometimes she gives it a very cruel bite. The poor little animals are very often seen with their skins pierced and lacerated in the most frightful manner. In this manner they go on until they have made their passage to whatever place she wishes to take her young one to." The males are described thus:—"One of a medium size will measure about 6 feet from nose to tail, and about 6 or 7 feet in circumference, and weigh about 5 hundredweight. They by far exceed these dimensions." "The fur and skin are superior to those of the female, being much thicker." On the neck and shoulders he has a thicker, longer, and much coarser coat of fur, which may almost be termed bristles; it is from 3 to 4 inches long, and can be ruffled up and made to stand erect at will, which is always done when they attack each other on shore or are surprised—sitting as a dog would do, with their head erect and looking towards the object of their surprise, and in this attitude they have all the appearance of a lion." "They begin to come into the bays in the month of October, and remain until the latter end of February, each one selecting and taking up his own particular beat in a great measure; but sometimes there are several about the same place, in which case they fight most furiously, never coming in contact with each other (either in or out of the water) without engaging in the most desperate combat, tearing large pieces of skin and flesh from each other; their skins are always full of wounds and scars, which however appear to heal very quickly." "At this place we saw hundreds of seals; both the shores and the water were literally swarming with them, both the tiger and black seal; but in general the tiger seals keep one side of the harbour, and the black seals, which are much the largest, the other side, but in one instance we saw a black and a tiger seal fighting. They were at it when we first saw them. We watched them about half-an-hour, and left them still hard at it; they fight as ferociously as dogs, and do not make the least noise, and with their large tusks they tear each other almost to pieces." "There is one seal which we all know particularly well wherever we see him; he appears to be the king of the mob which belong to Figure of Eight Island. He is a very large dark-coloured

c

bull of the tiger breed; we have named him Royal Tom. He is not at all afraid of us when we see him on shore; if the seals around him run away, Tom will not move, and takes very little notice of us. One day some of the men tried to drive Tom into the water, but he would not move for some time; but after some trouble I suppose they got him to start; he went leisurely down to the water, and there he remained scratching himself; Tom had a dry coat and did not fancy wetting it just then, and into the water he would not go." "In going up I found seal tracks nearly to the top of the mountain, which I reckon is about 4 miles from the water; and about 3 miles up I saw a seal." "We killed a cow and her calf this morning; we got milk from the cow after she was killed, which was very rich and good, and much better even than goat's milk." "The seals are very numerous here; they go roaring about the woods like wild cattle; indeed we expect they will come and storm the tent some night. We live chiefly on seal meat." "And a one-year-old seal, part of which we had roasted for dinner to-day—it was delicious." "One instance came especially under my notice of a cow, whose calf had been killed and taken away from her, roaming about the place where she lost it, incessantly bellowing, and without going into the water—consequently going without food for eight days. After the first few days her voice gradually became weaker, and at last could scarcely be heard. I made sure that she was dying. She survived it however, and on the eighth day went into the water."

Mr. Morris, in addition to the information already quoted in page 15, has kindly furnished me with the following interesting particulars of the history of the Southern Fur-Seal Fishery and the habits of the animal, which have the advantage of being derived from his own personal experience.

From him I learn the following particulars. The females in September come on shore to pup, and remain until about March. The pups are born black, but soon change to grey or silvery grey. The herd then go to sea for the remaining portion of the year, returning again in September with regularity.

During this absence at sea, the male pups have changed from the grey to a light brown colour, while the females remain unaltered.

In New South Wales the sealing trade was at its height from 1810 to 1820; the first systematic promoters of which were the Sydney firms of Cable, Lord, & Underwood; Riley & Jones; Birnie; and Hook & Campbell. The vessels employed by them were manned by crews of from twenty-five to twenty-eight men each, and were fitted out for a cruise of twelve months.

The mode of capture adopted was: the men selected for the shore party would number from six to eighteen, this being regulated by the more or less numerous gathering of the seals seen in the rookery. These men always land well to leeward, as the scent of the animal is very keen, and cautiously keep along the edge of the water, in order to cut off the possibility of retreat: then when abreast of the mob, they approach the seals and drive them up the beach to some convenient spot, as a small nook, or naturally formed inclosure: this accomplished, one or two men go in to the attack, while the others remain engaged in preventing outbreaks. As soon as a sufficient number have been slain to erect a wall of the dead, then all hands rush in to the general massacre.

To so great an extent was this indiscriminate killing carried, that in two years (1814, 1815) no less than 400,000 skins were obtained from Penantipod, or Antipodes Island, alone, and necessarily collected in so hasty a manner that very many of them were but imperfectly cured. The ship "Pegasus" took home 100,000 of these in bulk, and on her arrival in London, the skins, having heated during the voyage, had to be dug out of the hold, and were sold as manure—a sad and reckless waste of life.

Mr. Morris confirms Sir George Simpson and Mr. Musgrave in their account of the affection of the mother for her offspring: "At the time of the slaughter the female utters most piteous cries, alternately looking at you imploringly then at her young one"; such are his words.

ARCTOCEPHALUS GRAYII. Gray's Falkland Island Seal.

Synonym—*Arctocephalus Falklandicus.*—Gray, B. M. C. 1866, p. 55; Suppl. 1871, p. 25.

"Grey, under-fur red; young blackish; length 4 feet,[1]" "the fur very soft, elastic; the nose, cheeks, temples, throat, chest, sides, and underside of the body, yellowish white. It is easily known from all other fur-seals in the British Museum by the evenness, shortness, closeness, and elasticity of the fur, and the length of the under-fur. The fur is soft enough to wear as a rich fur without the removal of the longer hairs, which are always removed in the other fur-seals."[2]

This is clearly a species distinct from the common Southern fur-seal, and even from the two specimens of the Falkland Island sea-bear in the Edinburgh Museum, with whom it is compared; "t..e fur" of the latter "being considerably darker and harsher²"—distinctive qualities well and tersely defined.

The specific name *Falklandicus* having been appropriated almost by general consent for another animal, I beg to substitute that of *Grayii*.

ARCTOCEPHALUS EULOPHUS. Top-knot Seal of Patagonia.

Mr. Morris informs me that during his sealing voyages he occasionally met with a fur-seal, which he and those connected with him in the trade readily recognised as a distinct kind—by the diminutive size of the adult animal; by a top-knot of hair on the crown of the head; and by the soft, beautiful under-fur, unlike in colour to, and much more valuable for articles of ladies' wear than that of any other fur-seal they were in the habit of capturing.

[1] B. M. C., p. 55.
[2] Suppl., p. 26.

Relying upon the accuracy of observation acquired by long experience, I adopt this seal as a now species, and give the description of its appearance, as I received it from Mr. Morris. Length of the adult male about 4 feet, the female being smaller; the under-fur abundant, in texture fine, and very soft, and, when prime, of a rich plum-colour. Externally the animals are nearly uniform grey, and they possess a silver-grey tuft of hairs, or top-knot, on the head. This seal appears to be rare, only a few specimens having been taken; some were seen on the south-east coast of New Zealand, evidently stragglers driven far away from home. Mr. Morris has been told that they were formerly common on the shores of Patagonia and the Island of Juan Fernandez.

It will be found by comparing this seal with the Arctocephalus Grayii, or, Falklandicus of Dr. Gray, that in many important respects the two greatly assimilate, and that, in relation to others of the same group, dispersed over the Southern Seas, they in a marked manner alike disagree.

Dr. Gray, however, makes no allusion to the top-knot; but the skin of the adult female in the British Museum, upon which he based the specific characters, may possibly have been imperfect, or deprived of the coronal appendage during the process of preparation for export.

With such a doubt as to the presence or absence of so peculiar a feature in the London specimen, I cannot do otherwise than regard the present animal as a distinct species.

HAIR SEALS—*Adults*, with sparse under-fur.

Genus ZALOPHUS,[1] Gill.

Incisors $\frac{3\text{-}3}{2\text{-}2}$, canines $\frac{1\text{-}1}{1\text{-}1}$, molars $\frac{6\text{-}5}{5\text{-}5} = 34.$

Molars large, thick, closely approximated.

The skull, general form narrow; muzzle narrow; face considerably produced[2]; sagittal crest greatly developed.

In size intermediate between the Arctocephali and the Otariæ, and in external features distinct from either. The adult animals always possess a sparse under-fur.

ZALOPHUS GILLESPII, Macbain.   Californian Hair Seal.

Synonyms—*Arctocephalus Gilliespii*, Gray, B. M. C., p. 55.
  Zalophus Gillespii, Gray, Suppl., p. 28, Allen, B. M. C. Z., vol. 2, p. 68.
  Otaria Stelleri, Schlegel.

---

[1] Za, strongly, and λόφος, crest; in allusion to the great development of the sagittal crest of the skull.

[2] "In all the skulls we have of the genus (arctocephalus), a line drawn across the palate at the front edge of the zygomatic arch leaves one-third of the palate behind the line, and two-thirds in front of it; while in this species, Gilliespii, it leaves only one-fourth behind, and very nearly three-fourths in front of the line."—*Gray*, B. M. C., p. 55.

The male is larger in size and darker coloured than the female, which latter is described as being yellowish-brown on the sides, black along the back and head, and reddish-brown on the abdomen. The length of a skeleton in the Chicago Academy of Sciences, said to be that of a very old male, which I doubt exceedingly, would make the living animal somewhat over seven feet in length; a magnitude inferior to that of the Fur Seal, and greatly so to that of the aged male of the kindred species, the Z. lobatus of the Southern Hemisphere. This skeleton probably represents the remains of an aged female, for the dimension given above would correspond nearly with the size usually attained by our Counsellor Seal of similar sex. Mr. Allen "On the Eared Seals" distinctly states that "the two (Z. Gillespii & Z. lobatus) are nearly of the same size, and seem in general to have similar features." If so, the length acquired by a very old male of the Z. Gillespii would be nearer to eleven than seven feet.

Inhab.: Coasts of Japan and California.

In 1842 Schlegel described and figured in the Fauna Japonica, a hair seal found on the Japanese coasts, under the name of Otaria Stelleri, and in 1858 Dr. Macbain indicated a new species of seal, from the peculiarities exhibited in the form of a skull, which he obtained from California, now in the Museum of the Royal College of Surgeons, Edinburgh, naming it the Otaria Gillespii In the following year Dr. Gray redescribed Dr. Macbain's species, adding a figure, obtained from a cast of the skull. Seven or eight years subsequently, Dr. Peters of Berlin, by the examination of the specimens figured in the "Fauna Japonica," and comparing them with the Edinburgh one, felt assured that Schlegel's and Macbain's animals were of the same species.

About this latter period, Dr. Gill having seen other skulls of this species, all of which exhibited constant and radical differences between them and the forms found among the other eared-seals, very properly constituted the genus Zalophus.

ZALOPHUS LOBATUS, Gray. The Counsellor Seal.

Synonyms—*Otaria Cinerea*, Gray, in King's Narr. Australia.
 Arctocephalus lobatus, Gray, B. M. C. 1866, p. 50.
 Arctocephalus Australis, Quoy et Gaimard. Gray, B.M.C., p. 57.
 Zalophus lobatus, Allen, Bull. Mus. Comp. Zool., p. 44.
 Neophoca lobata, Gray, Suppl. 1871, p. 28.

The general colour of the adult male is black-brown, and that of the female a shade lighter. The pups are black, and covered abundantly with soft fur, which diminishes with age. Very old males will attain to twelve feet in length, but adults from eight to nine feet are usually met with. This species, formerly very abundant in Bass's Straits; N.W. coast of Australia; the Seal Rocks off Port Stephens, &c., is still found tolerably numerous; the commercial value of the animal consisting in the hide and oil only. Mr. Morris informs me he never saw this species at the Auckland Isles. In our Museum there are two stuffed specimens of very young animals, and it is much to be regretted that the adults, existing almost at our very doors, should remain unrepresented.

HAIR SEALS—*Adults*, without any under-fur.
Genus OTARIA, Péron.

Incisors $\frac{3-3}{2-2}$, canines $\frac{1-1}{1-1}$, molars $\frac{5-5}{5-5}$, or $\frac{6-6}{5-5}$ $=$ 34 or 36.

Upper outer incisors large, resembling canines; canines large, of the males extremely so; teeth of the female altogether much weaker and more sharply pointed than those of the male; cranium, subject to great individual variation, of the male broad, occipital portion elevated, which, in the very aged becomes immensely developed into crests; of the female, much narrower, and shallower, almost deficient of any occipital crest; mandible elongate, strong: limbs large, front feet with rudimentary nails; hinder, with the three middle nails long, the outer ones rudimentary; toe-flaps long; body clothed with hair, without any under-fur in the adults; males much larger than the females, and exhibit greater swimming powers, by possessing flippers proportionately much longer and stronger.

OTARIA STELLERI, Lesson. The Northern, or Steller's, Sea Lion.

Synonyms—*Leo marinus*, Steller, 1751.
   Otaria jubata, Péron, 1816.
   Otaria Stelleri, Lesson, 1828; Müller, 1841; Gray, B.M.C.,
    1850, p. 47, and 1866, p. 60; Sclater, P.Z.S., 1868,
    p. 190.
   Arctocephalus monteriensis, Gray, P.Z.S., 1859, and B.M.C.,
    1866, p. 49.
   Eumetopias Californianus, Gill, 1866.
   Eumetopias Stelleri, Peters, 1866; Gray, A. and M.N.H.,
    1866; Allen, Bull, Mus. Comp. Zool., vol. ii. p. 46;
    Gray, B.M.C.; Suppl. 1871, p. 46.

General colour: upper portion from pale yellow to reddish-brown, much darker towards the tail; under portion, dusky reddish-brown, darkest on the hinder part of the abdomen; frequently assuming a brindled appearance on some parts. Limbs, black-brown. The colour, however, varies much in different individuals—irrespective of age or sex.

The males attain to 13 feet in length, with a weight of from 1,500 to 1,800 lbs.: the females are more slender, and scarcely reach to one-fourth the weight of the male.

Inhabits the American coast, from California to Behring's Strait, and down the Asiatic coast to the Kurile Islands.

This species differs from the O. Jubata by having one pair less of upper molars, and in the modified form of certain portions of the cranium ; but in size, and general proportions, in the character of the hair, and its external colouring, both species bear a close resemblance.

From Mr. Allen's excellent treatise on the Eared Seals—from which the foregoing information has been derived—I extract notices by Captain Bryant and Mr. Lyman, of the habits of this Northern Sea Lion, which will be found to correspond greatly with those of the Southern animal. The former states :—

"The Sea Lion visits St. Paul's Island[1] in considerable numbers to rear its young "—" its habits are the same as those of the Fur-Seal "— " its skin is of considerable value as an article of commerce in the territory, it being used in making all kinds of boats, from a one-man canoe to a lighter of twenty tons' burden." "The rookery is on the north-east end of the island, and the animals have to be driven ten or eleven miles to the village to bring their skins to the drying-frames. It sometimes requires five days to make the journey, as at frequent intervals they have to be allowed to rest. It is a somewhat dangerous animal, and the men frequently get seriously hurt in driving and killing it. They are driven together in the same manner as the fur-seals are ; and while impeding each other by treading upon each other's flippers, the small ones are killed with lances, but the larger ones have to be shot.

"This animal is the most completely consumed of any on the island. Their flesh is preferred to that of the seal for drying for winter use. After the skins are taken off (2,000 of which are required annually to supply the trading-posts of the territory) they are spread in piles of twenty-five each, with the flesh side down, and left to heat, until the hair is loosened ; it is then scraped off, and the skins are stretched on frames to dry. The blubber is removed from the carcass for fuel, or oil, and the flesh is cut in strips and dried for winter use. The linings of their throats are saved and tanned for making the legs of boots and shoes, and the skin of the flippers is used for the soles. Their stomachs are turned, cleaned, and dried, and are used to put the oil in when boiled out. The intestines are dressed and sewed together into water-proof frocks, which are worn while hunting and fishing in the boats. The sinews of the back are dried and stripped, to make the thread with which to sew together the intestines and to fasten the skins to the canoe-frames."

Mr. Theodore Lyman observes :—"These rocks (Seal Rocks, near San Francisco) are beset with hundreds of these animals—some still, some moving, some on the land, and some in the water. As they approach to effect a landing, the head only appears decidedly above water. This is their familiar element, and they swim with great speed and ease, quite unmindful of the heavy surf, and of the breakers on the ledges. In landing they are apt to take advantage of a heavy wave which helps them to get the forward flippers on *terra firma*. As the wave

---

[1] Pribyloff group.

retreats, they begin to struggle up the steep rocks, twisting the body from side to side, with a clumsy worm-like motion, and thus alternately work their flippers into positions where they may force the body a little onward. It is quite astonishing to see how they will go up surfaces having even a greater inclination than 45°, and where a man would have to creep with much exertion." " In their onward path they are accompanied by the loud barking of all the seals they pass, and these cries may be heard at a great distance." "They play among themselves continually by rolling on each other and feigning to bite. Often, too, they will amuse themselves by pushing off those that are trying to land." " As they issue from the water, their fur is dark and shining, but, as it dries, it becomes of a yellowish brown. Then they appear to feel either too dry or too hot, for they move to the nearest point from which they may tumble into the sea. I saw many roll off a ledge at least twenty feet high, and fall like so many huge brown sacks into the water, dashing up showers of spray."

OTARIA JUBATA[1], Forster.   The Southern or Cook's Sea Lion.

Synonyms—*Phoca jubata*, Forster.
        Otaria jubata, Desmarest, Gray, Suppl. 1871, p. 13.
        Otaria leonina, Péron, Gray, B.M.C. 1866, p. 59; Peters, 1866.
        Arctocephalus Hookeri, ♀ ? Gray, B.C.M. 1866, p. 53.
        Phocarctos Hookeri, ♀ ? Gray, Suppl. 1871, p. 15.
        Sea Lion, Hamilton, Jard.: Nat. Libr., vol. 6, p. 237.

Inhabits Magellanic coast, Terra dell Fuego, Falkland and Auckland Islands, &c.

The external colouring of the hair, greatly diversified from birth to old age, exhibits not only the more permanent tinge acquired for the year after each shedding of the coat, but those intermediate changes which occur through the transitional state. It becomes, therefore, very difficult to recognize in full the many superficial tints, so as to suit the varying conditions of growth, but I may venture to offer the following summary of the colouration, collected from the best authorities at my disposal :—

The pups of both sexes, black-brown, or very deep chocolate ; nape of the neck and belly somewhat lighter ; under-fur very sparse and reddish, sensibly diminishing with age.

The male, young, above rich brown, beneath pale yellowish—adult, rich dark brown to brown grey ; beneath brownish or yellowish white ; mane with a brinded yellow and brown shade—aged, whitish grey.

The female, young and adult, light brown, grey, or dark grey ; abdomen, yellowish grey to light drab.

---

[1] Having a mane, maned.

It is recorded by Captain Cook and Mr. Forster that the largest
male seen by them was fourteen feet in length, with a weight of about
1,600 lbs., but such a size is now very rarely met with in this persecuted
race, and a male of nine or ten feet would, in the present day, be considered an animal of unusual bulk.

The head is small in proportion to the bulk of the body, and pug-
like in expression; the upper lip overhangs the lower one, and both
are furnished with long, coarse, black bristles; the body is thick and
cylindrical, more suited for rolling than walking. These animals herd
together in great numbers, but each group consists of one male and ten
or twelve adult females, with a family of from fifteen to twenty young
ones, from the sucking cub to the yearling.

According to Captain Cook—"it is not at all dangerous to go among
them, for they either fled or lay still. The only danger was in going
between them and the sea, for, if they took fright at any thing, they
would come down in such numbers that if you could not get out of
their way you would be run over. When we came suddenly upon
them, or waked them out of their sleep (for they are sluggish, sleepy
animals), they would raise up their heads, snort and snarl, and look
fierce, as if they meant to devour us; but, as we advanced upon them,
they always ran away, so that they are downright bullies." And Mr.
Forster, in his description of them, says:—"We put into a little cove,
under the shelter of some rocks, and fired at some of these fierce
animals, most of whom immediately threw themselves into the sea.
Some of the most unwieldy, however, kept their ground, and were
killed by our bullets. The noise which all the animals of this kind
made was various, and sometimes stunned our ears. The old males
snort and roar like mad bulls or lions; the females bleat exactly like
calves, and the young cubs like lambs."

SEALS without external ears.

Molars single-rooted.

Family II. TRICHECHIDÆ.[1]

WALRUS, or MORSE.[2]

Incisors $\frac{4\cdot4}{2\cdot2}$, canines $\frac{1\cdot1}{1\cdot1}$, molars $\frac{5\cdot5}{4\cdot4} = 34$ in young animal.
,, $\frac{0\cdot0}{0\cdot0}$, ,, $\frac{1\cdot1}{0\cdot0}$, ,, $\frac{4\cdot4}{4\cdot4} = 18$ in the old animal.

Upper canines prolonged downwards into enormous tusks; molars
small, slightly lobed, single-rooted; head round, obtuse; muzzle large,
very broad; lips thick, covered with coarse, strong, semi-transparent,
bristles; nostrils large, placed high on the muzzle; eyes small, prominent;

---

[1] θρίξ, τριχός, head with bristles, and ἔχω, having.
[2] *Morss*, Russian name of the animal.

mouth comparatively small; fore and hind limbs of same size; fingers of anterior feet nailless, the outer one longest; toes of posterior feet provided with small pointed nails,—the inner toes longest, the inner and outer toes lobed; tail very short, rudimentary; mammæ four, ventral; body exceedingly bulky, broadest round the chest; males and females of nearly the same size; hind limbs free, bent forward in repose; progression on land principally effected by the abdominal muscles; movements slow in water or on land; in habits sluggish, monogamous, and gregarious.

The Trichechus[1] Rosmarus,[2] Walrus,[2] Morse,[3] Sea-horse, Sea-cow, &c., is the only species known of this family—one so singularly conspicuous among the many strange forms which peculiarize the members of this order as to be instantly recognized. It is sufficient, then, without entering into comparative detail with other species, to allude to the two long and powerful canines, exserted from the upper jaw, and to the massive bones of the convex skull, modified wholly with reference to these enormously developed tusks, as forms of structure unpossessed by any other of the seal tibe, and consequently as being strikingly distinctive.

These tusks are sometimes two feet long, proportionately thick, and weigh between eight and ten pounds each.

They are much prized, for the ivory of which they are composed is denser and of a more permanent white than that of the elephant, and it, therefore, is more intrinsically valuable as an article of export.

The Walrus commonly measures from 10 to 14 feet in length; but an old male will reach to 20 feet, or even more, and in its bulk it corresponds to that of a large ox, although occasionally it will attain the dimensions of the elephant. The skin is thick, from one to two inches, black and smooth, and sparingly covered with short, stiff, hairs; in the adult, of a pale brown colour; in the young, blackish; and in the aged, of a whitish hue.

Inhabit.: The icy regions of the North.[4]

These animals appear to be omnivorous in their diet, for marine plants, molluscs, shrimps, cray-fish, and even portions of young seals, have been found in their stomachs.

When unmolested, they are quiet and inoffensive creatures, passing their existence harmoniously, in vast flocks upon rocky banks, or along sandy beaches, and during repose lie frequently huddled one over the other, like swine, delighting in solitudes far away from the haunts of man.

---

[1] θρὶξ, τριχός, head with bristles, and ἔχω, having.
[2] *Rosmar*, Norwegian name of Walrus, or *Hval-ros*, Whale-horse.
[3] *Morss*, Russian name of the animal.
[4] *Arctic and Antarctic Regions*, Wood, Illust. Nat. Hist., p. 515. *Polar Regions of both Hemispheres*, Sket. Nat. Hist., 1849, p. 199. *St. Lorenza, near Callao*, Bonelli, Trav. in Bolivia.

But when roused into anger in the defence of their young, or on being goaded with wounds, they then become formidable and dangerous enemies. Many interesting anecdotes are related of their strength and ferocity under these circumstances, but Captain Cook's short and expressive narative of their habits must for the present suffice. " The female will defend her young one to the very last, and at the expense of her own life, whether in the water or on the ice; nor will the young one quit the dam, though she be dead; so that if you kill one, you are sure of the other." Again, "the female in particular, whose young had been destroyed and taken into the boat became so enraged, that she attacked the cutter and stuck her tusks through the bottom of it."

The flesh is highly valued, and greedily eaten by the natives; the skin being thick and tough, is useful for many purposes;—in ancient times, when cut into strips and plaited, it formed the ropes and cables for the vessels of northern countries, and the finer portions made into lines, were used for the capture of whales. In modern times the skins are sent to America and England, and manufactured into carriage traces and other harness, or rendered down into glue; the oil, although not abundant, is superior in quality; but the teeth constitute the most valuable product of this animal, for the ivory being of a beautiful texture, and capable of retaining its whiteness, is extensively used—by the Chinese, for the wonderful knick-knacks and other curiosities they produce from the lathe—and by Europeans, for the supply of artificial teeth, and many kinds of ornamental work dependent on these properties.

Family III. CYSTOPHORIDÆ.[1]

SEA ELEPHANT: HOODED SEAL.

Incisors $\frac{2 \cdot 2}{1 \cdot 1}$, canines $\frac{1 \cdot 1}{1 \cdot 1}$, molars $\frac{5 \cdot 5}{5 \cdot 5} = 30$.

Outer incisors large, formed like the canines; molars with small compressed crowns and greatly swollen single roots; head short, broad; muzzle of the males furnished with a dilatable bladder-like appendage; whiskers, long, thickish, waved, obtuse at their tips; nostrils large; eyes large, prominent; nails elongated, pointed, obsolete in the hinder feet of the macrorhinus; tail very short. The animals during progression on land move principally by means of the abdominal muscles and extremely flexible spine, assisted materially by their flippers. In repose the hind limbs are stretched backwards in a line with the body.

In habits polygamous and gregarious.

Genus MACRORHINUS, F. Cuvier. Sea Elephants.

The adult males possess the power of elongating the nose into a tubular proboscis, resembling somewhat the proboscis of the elephant. In the female this dilatable appendage is undeveloped. Forehead,

---

[1] κύστις, a bladder, and φέρω I bear. The generic name Cystophora is likewise applied to the species of a marine plant, allied to the gulf-weed.

convex; hairs of the whiskers very long, large, roundish, and slightly
waved; similar hairs in tufts over each eye and on each cheek; fore
feet with longish claws, the first one being the smallest; hind feet with
the outer toes large, the three middle ones small, all of them without
nails; eyes large and prominent.

MACROBHINUS[1] ELEPHANTINUS,[2] Molina.

Synonyms—*Phoca elephantina*, Molina.
Phoca proboscidea, Péron, Cuvier, Hamilton.
Macrorhinus proboscideus, F. Cuvier, 1827.
Mirounga proboscidea and *Ansonii*, Gray, 1827.
Cystophora proboscidea, Nilsson, 1837.
Morunga elephantina, Gray, B.M.C. S. & W., 1866, p. 38,
   Suppl. 1871, p. 4.
Miourounga. Australian Aboriginals.

The sea elephants retain the semi-terrestrial habits of the eared seals,
although they differ so materially from them in the structure of the
posterior extremities, these becoming so confined within the integuments
of the body as to possess but little or no power of motion.

This cramped condition of the hinder limbs, common to the whole of
the seal tribe, with the exception of the eared-seals and the walrus, is
in this species slightly mitigated by the thick and stout form of the
pelvic bones, which permits a freer use of these members, and by the
greater expansion and stoutness of the shoulder-blades, which strengthen
the flippers, and render their assistance more effective; so that a power
of locomotion on land is attained of an intermediate character between
that exhibited by the Otariadæ and Phocidæ.

The sea elephants were formerly found in great abundance inhabiting
many of the numerous islands lying between the thirtieth degree of
south latitude even to the verge of the antarctic circle, as Juan Fernandez, Staten Island, Falkland Islands, South Georgia, Tristan
d'Acunha, Kerguelen's Land, and other spots, where sandy beaches and
fresh-water swamps exist: a geographic range so vast, as to comprise
at least two-thirds of the whole area of that portion of the Southern
Seas comprised within the latitudinal belt above specified. Their
powers of locomotion, however, are so great, that frequent stragglers
have been captured on the coasts of Australia.

In fact, it appears that this animal affords another illustration of the
extensive habitat originally enjoyed by certain species, such as the Sea
Lions and the Seals of Commerce of the North and the South, until
driven away by continuous persecution to seek the more restricted and
outlying homes adequate for their reduced numbers.

---

[1] μακρός, long, and ῥίν, the nose.
[2] of an elephant.

This enormous creature, commonly twenty feet in length, is not only by far the largest among the seal-tribe, but the aged male, acquiring a length of thirty feet and a girth of twenty, will double the dimensions of the great elephant itself. This huge mass, tremulous with fat, and the peculiar prolongation of the nostrils when the animal is excited, to a foot beyond the lips, are of themselves sufficiently characteristic; but to complete the description it may be added, that in colour the males are usually greyish, or bluish grey, rarely blackish-brown, and that the females are olive brown above, passing into yellowish-bay beneath, and very much smaller in size in every admeasurement of the body.

In disposition these seals are mild, fearless, and apathetic, avoiding intercourse with man by tenanting the most desolate shores, and when intruded upon by him, showing no resentment, so that he may walk amongst them without fear, and even bathe in their midst without risk of injury, so long as they are left in peace to indulge in a slothful existence, spent principally in wallowing in the mud or reposing among the long grass, which luxuries they enjoy in social herds of thousands, until the general harmony is temporarily disturbed by the advent of the season of courtship, when, madly roused, the males are urged into desperate conflicts with each other.

"These seals," observes Captain Carmichael, in his description of the island of Tristan d'Acunha, "pass the greater part of their time on shore; they may be seen in hundreds lying asleep along the sandy beach, or among the long grass which borders the sea-shore.

"These huge animals are so little apprehensive of danger that they must be kicked or pelted with stones before they make any effort to move out of one's way. When roused from their slumber, they raise the fore part of their body, open wide their mouth, and display a formidable set of tusks, but never attempt to bite. Should this, however, fail to intimidate their disturbers, they set themselves at length in motion, and make for the water, but with such deliberation, that on an excursion we once made to the opposite side of the island, two of our party were tempted to ride upon the back of one of them and rode him fairly into the water."

It is said by another authority, that when the females produce their young, the males form a line between them and the sea to prevent the desertion of their charge, even for the shortest space of time. This period of nursing and imprisonment lasts for seven or eight weeks, during which time the females are debarred from food, and become extremely emaciated.

When taken young, they are easily tamed and become very affectionate: one petted by an English sailor became so attached to his master, from kind treatment for a few months, that it would come at his call, allow him to mount upon his back and put his hands into its mouth.

The fishers use, in order to kill them, a lance twelve or fifteen feet long, with a sharp iron point of about two feet. With great address they seize the moment when the animal raises his fore-paw to advance, and plunging the weapon to the heart, he immediately falls down drenched in blood. The females rarely offer the least opposition when attacked, but they endeavour to fly; if prevented, their countenance assumes the expression of despair, and they weep piteously.[1] "I have myself," says M. Péron, "seen a young female shed tears[1] abundantly, whilst one of our wicked and cruel sailors amused himself at the sight, knocking out her teeth with an oar, whenever she opened her mouth. The poor animal might have softened a heart of stone, its mouth streaming with blood and its eyes with tears."

The Elephant Seal is valued on account of the oil which it yields in abundance, an adult male averaging seventy gallons, and which in quality is limpid, free from smell, never becoming rancid, and in burning, smokeless. The hide, also, is from its strength and thickness, extensively used for carriage and horse harness.

New Georgia alone formerly supplied the English market with twenty thousand gallons of oil annually; this article being from its quality greatly adapted for softening wool, and for other purposes in the manufacture of cloth.

The food of this animal appears to consist principally of cuttle-fish and sea-weed.

MACRORHINUS ANGUSTIROSTRIS.[2] Gill. Californian Sea Elephant.

Synonym—*Morunga angustirostris*, Gray, Supp. B.M.C. S. & W. 1871, p. 5.

This species, introduced recently to our notice by Dr. Theodore Gill, of Washington, is the northern representative of the Sea-Elephants, so long and so well known in our hemisphere.

Both kinds appear to be equally bulky, but differ principally in the narrower muzzle possessed by the American animal.

Inhabits California, from Cape San Lucas to Point Reyes.

Genus CYSTOPHORA, Gray.

Adult males possess a dilatable globular sac-like appendage upon the crown of the head, immediately connected by a cartilaginous crest with the nostrils, and which, by their agency, can be distended or collapsed at will. In the females and the young this singular hood is rudimentary, scarcely perceptible. Muzzle very broad, hairy; hairs of the whiskers long, whitish, waved, compressed at their base; those of the body long and coarse, with an under-fur, short, soft, and thick; the limbs are all distinctly clawed.

---

[1] Evidently a common and natural habit, quite irrespective of anguish, for "when on shore, they" (the fur seals of the Auckland Islands) "appear to be constantly weeping."—*Captain Musgrave's Narrative.*

[2] *angustus*, narrow, and *rostrum*, beak or snout.

CYSTOPHORA CRISTATA. *Erxlehen.* Crested or Hooded Seal.

Synonyms—*Phoca cristata,* Erxl.
 Phoca mitrata, Cuvier.
 Cystophora cristata, Nilsson. Gray, B.M.C., 1866, p. 41.
 Stemmatopus cristatus, F. Cuvier.
 Stemmatopus mitratus, Gray.
 Phoca leonina, Linnæus.
 Hooded Seal, Pennaut.
 Bladder-nose of Sealers.

This seal is from 8 to 12 feet in length, and in the different stages of growth varies considerably in colour, which in the full-grown male is dark brown, approaching to black, relieved by numerous largish, irregularly shaped rings of a greyish hue, scattered over the body. The young are much lighter coloured, from grey to brown-grey along the back, with the abdominal portion white.

Speculative writers have ascribed to this bladder-like appendage many uses to which it can be beneficially applied—such as, that, being connected with the nostrils, it is subsidiary to the sense of smell,—that it is a reservoir for air, to be consumed when under water,—that the head can be buried in it, as in a monk's hood,—and that it can be drawn over the eyes, like a cap, to defend them against the storms, waves, stones, and sand. I need scarcely say that to attribute such properties to this peculiar sac is purely hypothetic; for the young males—even up to three years old—and all the females, exhibit this peculiarity only in a rudimentary state, yet have their faculties in or out of water as keenly developed and as well protected from those injuries to which they are alike exposed. This dilatable globular appendage on the top of the head, however, clearly indicates—in this species—the puberal maturity of the male, and serves to modulate the voice, and to give it those inflections of tones so highly expressive of desire or rage.

Inhabits the North Atlantic.

This animal, the Harp, the Ringed and the Common Seal, are objects of extensive capture chiefly for the sake of their skins, which, by careful preparation, can be applied to many useful and ornamental purposes.

By the natives of Greenland every portion is converted to some valuable use: the flesh, the oil, and the blood, are greedily consumed; their houses are covered with, and their boats made of, the skins of the older animals; and these are firmly sewn together by the strong fibres of the sinews; while the soft fur furnishes various articles of apparel; the stomachs are converted into fishing buoys; the semi-transparent internal membrane provides the substitute of glass for their windows; and the teeth form their spear-heads.

The Esquimaux seal hunter, in taking any of these animals, proceeds thus: "Having ascertained by an examination of the ice that a seal is near at hand (and he can discover this by the small hole left by the animal to enable him to raise his head above the water to breathe) he sets to work to form a kind of arm-chair of square lumps of ice, the back of course placed to the windward, when, resting his spear, to which a long line is attached, on a small piece of ice, so that he may lift it with the least possible noise, he places himself in this comfortless seat, and patiently awaits, perhaps for hours, the return of the animal to his blow-hole."

In their domestic habits the Crested Seals resemble the other polygamous groups, existing at certain times in comparative harmony with their neighbours, at others, the whole community becomes involved in strife.

Molars, more or less, with double roots.

Family IV. PHOCIDÆ.[1]

Monk Seal, Common Seal, Grey Seal, Sea Leopard, &c.

Hind limbs, when at rest on land, are directed backwards nearly in a line with the body, by the integuments of which they are so enveloped and confined as to possess but little or no power of motion, the feet been capable of moving only in an obliquely lateral direction. The progression on land, therefore, like that of the preceding family, but more restricted, is effected by means of the abdominal muscles and extremely flexible spine, assisted materially by the front limbs. In many other important portions of their structure, they likewise differ greatly from the Eared Seals:—the skull is but moderately crested; the shoulder-blade is reduced in size; the pelvis is comparatively small, and in its form exhibits no unusual sexual difference, being alike broad in both sexes; and the pelvic bones are thin and slender. The hair which thickly clothes the body is short, closely pressed against the skin, more or less soft and woolly, and extensively used in the manufacture of articles of wear, although greatly deficient in quality to the under-fur which distinguishes the Fur Seal of commerce.

Genus MONACHUS,[2] Fleming.

Synonyms—*Pelagios*,[3] F. Cuvier, Gray.
Pelagius,[3] F. Cuvier, Fischer, Nilsson.

Incisors $\frac{2\cdot2}{2\cdot2}$, canines $\frac{1\cdot1}{1\cdot1}$, molars $\frac{6\cdot5}{5\cdot6} = 32$.

Upper incisors indented transversely at their edge, so that the lower ones, when the mouth is closed, fill up these indentations. Molars crowded, obtusely pointed, slightly lobed; anterior one of each jaw

---

[1] From φώκη, *phoca*, a seal.
[2] A monk.
[3] That lives in the sea.

with a single root, the others with two roots; claws of fore feet rather
flat, of hinder ones conical, very small; whiskers stiff, short, smooth;
muzzle with a slight central groove. The crests of the skull, and the
expansion of the shoulder-blade, are more strongly developed in this
genus than in the following ones.

MONACHUS[1] ALBIVENTER,[3] *Boddaert.* Monk Seal.

Synonyms—*Phoque à ventre blanc,* Buffon, Cuvier.
Phoca monachus,[1] Herm, Desmarest.
Pelagios[2] *monachus,*[1] F. Cuvier, Blainville, Owen.
Pelagius[2] *monachus,*[1] Nilsson, Gray.
Phoca albiventer,[3] Boddaert.
Mediterranean Seal, Shaw.
Monachus albiventer, Gray, B. M. C. 1866, p. 19.

The well-known sub-tropical Seal has been long considered as the
only representative of the genus, and I believe it is still so regarded by
the generality of writers on the animals of this order; although
Dr. Gray, in the Catalogue and Supplement, so continually referred to,
has introduced another, a tropical species,[4] from the distinctive
characters afforded by an imperfect skin, which the British Museum
received from Jamaica. It is to be understood, however, that the
foregoing generic characteristics are derived from the Monk, or White-
bellied Seal of the Mediterranean. Inhabits both shores of and the
islands in the Mediterranean, and said occasionally to be found at
Madeira and the Canaries.

Fully-grown adults are from 10 to 12 feet in length. Monk Seals
have upon several occasions been partially domesticated, and thus
opportunities have been given for ascertaining many of their ways and
disposition. One, a few years ago, was exhibited in London as the
"Talking Fish"; at the word of command, it would utter various
sounds, from a bark to the hoarse bellow of a bull, would offer its lips
to be kissed, and perform within its tub many pleasing feats of agility.

Buffon, F. Cuvier, and other eminent men, have at different times
availed themselves of similar chances for observation, and fortunately
have recorded minutely their experience. The former naturalist, in
describing a male taken in the Adriatic, remarks: "The white-bellied
Seal we saw in December, 1778. Its aspect is mild, and its disposition
not fierce; its eyes are quick and intelligent, or, at all events, they express the sentiments of affection and attachment to its master, whom it
obeys with the utmost readiness. At his order we have seen it lay
down its head, turn in various directions, roll round and round, raise
the fore-parts of its body quite erect in its trough, and shake hands

---

[1] A monk.
[2] That lives in the sea.
[3] *Albus,* white, and *renter* belly.
[4] B. M. C. Seals and Whales, 1866, p. 20. Monachus tropicalis.

D

with him. Previous to being tamed, it bit its master furiously when interfered with, but when subdued, it became quite mild, so that it could be handled with all freedom. You might thrust the hand into its mouth, and rest your head on that of the Seal. When its master called, it answered, however distant it might be ; it looked round for him when it did not see him, and on discovering him after an absence of a few minutes, never failed to testify joy by a loud murmur. Some of its accents were sweet and expressive, and seemed the language of pleasure and delight.

" The period between its several inspirations was very long, and in the interval the nostrils were accurately closed, during which time they appeared like two longitudinal slits on the end of the snout. The creature opened them to make a strong expiration, which was immediately followed by an inspiration, after which it closed them as before ; and often allowed two minutes to intervene without taking another breath. The breathing was accompanied with a loud snuffling noise. When drowsy, it did not promptly attend to its master, and it was only by putting food under its very nose that it could be excited to its accustomed energy and vivacity. It then raised its head and the upper part of its body, supporting itself on its fore-paws, to the height of the hand which held the fish ; for it was scarcely satisfied with any other aliment, having a preference for carp, and still more for eels ; these, though raw, were seasoned to its taste by rolling them in salt. It required about 30 ℔s. of these live fish every day ; it greedily swallowed the eels entire, and even the carp which were first offered it, but, after devouring two or three entire, it subjected them to some preparation by crushing their heads with its teeth, then partially gutting them, and concluded by gulping them head-foremost. This individual was 7½ feet long ; its skin was covered with a short, smooth, shining hair of a brown colour, mixed with grey principally upon the neck and head, where it was spotted ; the fur was thicker on the back and side than on the belly, where there was a large white marking, which mounted up upon the flanks. The nostrils were neither inclined nor were they placed as in terrestrial quadrupeds, but extended vertically on the extremity of the snout. The eyes were large, full, of a brown colour, and like those of an ox."

M. F. Cuvier furnishes in 1813 a detailed description of a female Seal which was captured in 1811 ; from which memoir the following short extracts are taken. " For two years it has been kept in a trough, which scarcely exceeds its own dimensions, being only one foot longer, and two feet broader than itself. It every day receives several pounds of fresh-water fish, and usually spends nine or ten consecutive hours in water ten inches deep. At the close of the day the water is removed, that the animal may be dry during the night, and, in spite of this artificial mode of life, it enjoys excellent health.

" The length of this animal is between seven and eight feet, and the general form is very like that of the common Seal. Its colour in the water is black on the head, back, tail, and upper part of the feet, whilst the chest, sides, and belly, and the under portion of the neck, tail, paws, and sides of the head, are of a yellowish light grey. When it is dry, the black portions are not so deeply coloured, and the white parts are more yellow. The skin is everywhere of a slaty colour. The tail is three inches long, and without movement; the eyes are large, and the cornea is very flat in comparison with other quadrupeds; two hairs, similar to those of the lip, are seen above each eye; the pupil exactly resembles that of the domestic cat; the nostrils are naturally closed, and open only at the will of the animal; the ear has no trace of an external auricle; the orifice of the auditory canal is situated nearly opposite the tympanum. It sleeps during the live-long night, and cannot be kept awake during the day without the most unceasing perseverance. During sleep it is often observed covered with the water at the bottom of its trough, where of course it cannot breathe, and there it continues for an hour at a time."

The habits of the animals of this species in a state of nature are similar to those of the crested Seals.

Genus Phoca, Linnæus.

Incisors $\frac{3\text{-}3}{2\text{-}2}$, canines $\frac{1\text{-}1}{1\text{-}1}$, molars $\frac{6\text{-}5}{5\text{-}5} = 34$.

Incisor small, pointed; molars, placed in an oblique position along the jaws, moderately large-lobed, somewhat crowded; anterior one of each jaw with a single root, all the others double-rooted; teeth of moderate size; whiskers small, waved; muzzle with a distinct central groove; fingers gradually shortening from the first to the inner one, the Leporine Seal excepted; toes, inner and outer ones large, long, the middle ones shorter; claws large, conical, sharp; habits similar to the preceding genus.

Phoca vitulina,[2] Linnæus.   Common Seal.

Synonyms—*Callocephalus*[1] *vitulinus*.[2]—F. Cuvier, Gray, B. M. C., p. 20.
Common Seal, Pennant.
Phoque commune, Buffon.
Sea-Calf, or *Sea-Dog* of Sailors.
Meerhund, Zeehund, Seelhund, of the Germans, Dutch, and Danes.

The Common Seal furnishes another example to those previously given of the wide geographic range enjoyed by many animals of the same species. By means of well authenticated specimens, it is ascertained to inhabit nearly every coast washed by the cold waters of the Northern Seas, and it is moreover found in the salt sea of the isolated Caspian, and, far distant from the ocean, in the fresh waters of the

---

[1] καλός, beautiful, and κεφαλή, the head.
[2] *Vitulinus*, calf-like.

Lake Baikal; that is to say, in general terms, from Greenland eastward into Eastern Siberia. This nearly circumpolar belt of occupation is fully as extensive and infinitively more difficult to comprehend than the extensive habitat I attribute to the Arctocephalus Falklandicus, and would afford more reasonable grounds to the supporters of the theory of limited location,[1] for the separation into species, if not into genera, of the several examples of this familiar animal.

The colour of the Common Seal, is on the upper portion of the body, yellowish-brown of various shades, but commonly dark, and frequently mottled, or spotted over with darker: beneath much paler, yellowish-white. The usual length is from four to six feet, but an aged male will exceed these dimensions, and it has been known to weigh two hundred and twenty-four pounds.

Inhabits Greenland, the North Sea, the Baltic, the Caspian Sea, and Lake Baikal. It is still found in considerable numbers on the English, Scottish, and Irish Coasts.

To describe all of the ascertained habits of this seal in the state of nature, or when semi-domesticated, would amount in many instances simply to the repetition of anecdotes already given of other members of the family; but I will offer a few additional descriptive extracts of manners and disposition, which, although derived exclusively from this species, will tend materially towards perfecting our knowledge of a group so wonderfully similar in every important feature of organization.

Professor Trail relates that, "A young seal, about two and a half feet long, would suck one's fingers readily, was greedily fond of milk, and seemed a social animal. When thrown into the sea, it speedily returned to the shore, and made back for its favourite position, the kitchen hearth, the stone of which was elevated about four inches above the floor, and it generally laid itself so close to the embers of a peat fire burning there, that it often singed its fur." [2] "One in particular became so tame that he lay along the fire among the dogs, bathed in the sea, and returned to the house: but having found his way to the byres[3], used to steal there and suck the cows; on this account he was discharged and sent to his native element." [4] "During a residence of some years in one of the Hebrides, I had many opportunities of witnessing this peculiarity (partiality for musical and other sounds), and, in fact, could call forth its manifestation at pleasure. In walking along the shore in the calm of a summer afternoon, a few notes of my flute would bring half a score of them within thirty or forty yards of me, and there they would swim about, with their heads above water, like so many black dogs, evidently delighted with the sounds. For half-an-hour, or indeed, for any length of time I chose, I could fix them to the spot; and when

---

[1] "We now know that the species (Otariadæ) have a very limited geographical distribution." Gray. Suppl. 1871, S. & W. p. 7.
[2] Naturalist's Library, p. 134.
[3] *Byre*, Scotch, a cow-house.
[4] Mr. L. Edmonstone.

I moved along the water's edge, they would follow me with eagerness, like the dolphins who, it is said, attended Arion, as if anxious to prolong their enjoyment. I have frequently witnessed the same effect when out on a boat excursion. The sound of a flute, or of a common fife, blown by one of the boatmen, was no sooner heard, than half-a-dozen would start up within a few yards, whirling round us as long as the music played, and disappearing one after another when it ceased.[1] "

" The church of Hoy, in Orkney, is situated in a small sandy bay, much frequented by these creatures; and I observed, when the bell rang for divine service, all the seals within hearing swam directly to the shore, and kept looking about them, as if surprised, rather than frightened, and in this manner continued to wonder as long as the bell rang." " Whilst I and my pupils," says Mr. Dunbar, " were bathing, as was our custom, in the bosom of a beautiful bay, named Seal Bay, in Orkney, numbers of these creatures invariably made their appearance, especially if the weather was calm and sunny, and the sea smooth, crowding us at the distance of a few yards, and looking as if they had some kind of notion that we were of the same species, or at least, genus, with themselves.

" The gambols in the water of my playful companions, and their noise and merriment, seemed to our imagination to excite the seals, and to make them course around us with greater rapidity and animation. At the same time, the slightest attempt on our part to act on the offensive, by throwing at them a stone or shell, was the signal for their instantaneous disappearance; each, as it vanished, leaving the surface of the waters beautifully figured with a wavy succession of concentric circles."

PHOCA FŒTIDA[2], Müller.   The Ringed Seal.

Synonyms—*Phoca fœtida*[2], Müller.
    Phoca hispida[3], Erxleben—O. Fabricius.
    Phoca fasciata[4], Shaw.
    Phoca annellata[5], Nilsson.
    Callocephalus hispidus[3], F. Cuvier.
    Pagomys fœtidus[2], Gray, B. M. C., S. and W. 1866, p. 23.

Inhabits Greenland—North Sea—Lake Baikal.

This species is about the size and build of the preceding one, but is readily distinguished by the marbled disposition of the colouring of the hair on the upper portions of the body, which appearance is caused by numerous whitish ovate ocellated spots, about two inches long, distributed over a brown ground colour, darkest along the back, and paling beneath to nearly white. The young are of a darker hue, and their skins are not relieved by the annular spots.

---

[1] Mr. Lizars.
[2] *Fœtidus*, stinking, rank.
[3] *Hispidus*, rough, shaggy.
[4] *Fasciatus*, banded.
[5] *Annellatus*, with little rings.

Old males are said to acquire a disgusting smell, from which un-enviable circumstance the Latin specific name has been derived.

PHOCA GRŒNLANDICA[1], Müller.   The Harp Seal.

Synonyms—*Phoca Grœnlandica*, Müller.   O. Fabricius.
   Phoca oceanica[2], Lepechin Fischer.   Jardine's Nat. Lib.
   Callocephalus grœnlandicus.   F. Cuvier.
   Pagophilus grœnlandicus.   Gray, B. M. C. 1866, p. 25.
   Harp-Seal, Pennant, Bell, Hamilton.

Inhabits Greenland, North Sea.

" Until six or seven weeks old, white, called *white coats* at Newfound-land ; at one year old they have small spots ; at two years old they have large spots, and the males are called Lampiers ; at three years old the males and females have the harp-shaped band, and are then called *saddle-backs.*"[3]

The fur of the adult is greyish-white, the back being marked by a blackish horse-shoe-shaped band, arching backwards from the shoulder to within a few inches of the tail.   This band is broad at the sides, while its outline is very irregular ; the anterior half of the head exhibits the same deep brownish-black colour of the band.

The Harp Seal is very abundant in the deep bays and mouths of rivers along the coast of Greenland, living among the floating masses of ice, and preying principally upon the Arctic salmon and other fish—and occasionally upon molluscs.   In size and general make it resembles the two preceding animals, but its fur and oil are alleged to be of better quality than theirs.

PHOCA BARBATA,[4] O. Fabricius.   The Leporine Seal.[5]

Synonyms—*Phoca barbata*, O. Fabricius ; Müller, Nilsson, Fischer, &c.
   Phoca leporina,[5] Lepechin.
   Callocephalus leporinus, F. Cuvier.
   Phoca barbata, Gray, B. M. C., 1866, p. 31.
   Leporine Seal, Pennant.

This Seal and the following one are frequently mistaken for each other, for they bear a general external resemblance, are similar in size, being by far the largest of the species which I have attached to this family, and both are found on the British and Irish Coasts.   Their structural characters and habits, however, vary so much as to render them palpably distinct.

The ordinary length of the adult animal may be taken at about nine feet, but the aged will reach to twelve feet, or even more.   In colour,

---

[1] Greenland.
[2] Oceanic.
[3] Jukes, Newfoundland.
[4] Bearded.
[5] Hare-like.   This species is likewise known as the Great Seal, the Great Bearded Seal, the Hare-like Seal.

the male is brownish-black, fading into yellowish on the abdominal parts ; the young are much lighter in hue, which assumes a greenish cast. The females are similarly coloured, but the underneath portion is greyish.

This species differs from the preceding ones, in having the central finger the longest, and the outer and inner ones the shortest.

Inhabits the Northern Seas, and occasionally found on the Scottish and Northern Coasts of England.

Genus HALICHŒRUS,[1] Nilsson.

Incisors $\frac{3\cdot3}{2\cdot2}$, canines $\frac{1\cdot1}{1\cdot1}$, molars, $\frac{5\cdot5}{5\cdot5} = 34$.

Canines moderate in size ; molars conical ; upper ones simple ; lower ones slightly lobed ; the two posterior ones on each side of the upper jaw, and the posterior one of the lower, are double-rooted ; the remaining ones with single roots : head very flat ; bones of the face strongly developed : brain comparatively very small ; muzzle simple, broad, rounded, truncated ; whiskers notched at their edges ; claws conical, elongated, sharp.

Habits very moderately gregarious ; scarcely susceptible of domestication.

HALICHŒRUS[1] GRYPUS,[2] O. Fabricius. The Grey Seal.

Synonyms—*Phoca grypus*, O. Fabricius.
Halichœrus grypus, Nilsson, Gray, B. M. C., 1866, p. 34.
Grey Seal, Bell, Brit : Quad.

Dr. Ball, of Dublin, in his excellent account of the habits of the Grey Seal, remarks that its colour varies so much from sex, age, season, &c., that it cannot be regarded of value as a specific character ; which observation, as I have before pointed out, is equally applicable to many species other than the animal he is describing. It is, however, readily distinguished by the more permanent characters of a straight profile, fierce aspect, and greater proportionate length of body to the rotundity.

In its habits it is usually solitary, associating only in small parties, and in its disposition devoid of that intelligence and mildness so strikingly conspicuous in others of its kind.

"My father," writes Dr. Ball, "has made several attempts to rear and tame this seal, but in vain. It appears scarcely susceptible of domestication, and the development of the skull seems to indicate as much ; for the size of the brain of a specimen nearly eight feet long did not exceed that of one of the common seal (Ph. vitulina) of less than four." To which convincing fact, Mr. Bell, in the 1st volume of his "British Quadrupeds," adds, "It is impossible not to be forcibly struck with the contrast between the cerebral development of this genus and that of the former, and the relation between the difference

---

[1] ἅλς, the sea, χοῖρος, hog or pig.
[2] γρυπός, having the beak hooked.

of structure and their susceptibility of domestication. It is exactly analogous to the distinction between the crania of baboons and those of the higher groups of quadrumanous animals."

In colour the very young female is of a dull yellowish-white, which in a month or six weeks becomes variously blotched with grey; as the animal advances in age, these blotches almost disappear on the upper part of the body, but they remain very distinct on the lower part and on the breast. From a peculiarity in the hair of the adult, it being considerably recurved, and as if its upper surface were scraped flat with a knife, the animal, when dry, and with its head turned towards the spectator, appears of a uniform silver grey, whilst viewed in the opposite direction it appears altogether of a sooty brown colour, the spots or blotches being only visible on a side view. The young male has long yellowish hair, slightly tinged with brownish-black along the back.

The grey seal will sometimes attain a length of twelve feet, and a weight of 650 lbs., but such large specimens are seldom encountered. Nilsson states that in the Baltic it is a solitary animal, but on the coasts of Ireland, where it is still numerous, and on those of Scotland, this species is unquestionably gregarious, associating in small families of from ten to fourteen members.

Genus STENORHYNCHUS,[1] F. Cuvier.

Incisors $\frac{2\text{-}2}{2\text{-}2}$, canine, $\frac{1\text{-}1}{1\text{-}1}$, molars, $\frac{5\text{-}5}{5\text{-}5}$ = 32.

Incisors conical, the outer upper ones large, resembling canines, one species excepted; molars distinctly trilobate; anterior one in each ramus single-rooted; the others with two roots; muzzle simple, hairy between and above the nostrils; whiskers small, wavy, tapering; claws of fore feet small, of hind feet obsolete, or nearly so.

STENORHYNCHUS[1] LEPTONYX,[2] Blainville. The Sea Leopard.

Synonyms—*Phoca leptonyx*, Blainville.
　　　　　Stenorhynchus leptonyx, F. Cuvier; Gray, B. M. C. 1866, p. 16.
　　　　　The small-nailed Seal (?), Jardine, Nat. Libry., p. 180, pl. 11.
　　　　　The Leopard Seal (?), Jardine, Nat. Libr., p. 183, pl. 12.

There are two stuffed specimens of this species in the Australian Museum; one, recently obtained, is admirably set up, the various admeasurements being taken from the animal when living, the other but indifferently. These afford another example to the many, that colour, and variations of marking, when considered alone, are but unreliable evidence in distinguishing species, for on these points they differ considerably. Their skulls, however, allowing for those minor

---

[1] στενός, narrow, and ῥύγχος, the beak.
[2] λεπτός, slender, and ὄνυξ, the nail.

deviations commonly attendant upon age, and individual peculiarities, are too alike in their general structure to permit of any doubt as to their specific identity.

Another perfect skull from Lyttleton, New Zealand, presented to this institution by Dr. Schutte, and a water-colour drawing, by Mr. Angas, of a sea leopard captured at Newcastle, exhibit further slight differences in the dentition or in the external colouring of the hair, yet they and the two before-mentioned animals are unquestionably of the one and the same species.

I have been partly led to offer these observations, because the colouration and general outline of the sea leopard (Phoca leopardina) figured in Plate 12 of Jardine's Naturalist's Library resemble those of the aged male in our Museum; while the small-nailed seal (Phoca leptonyx) represented in Plate 11, approaches more in its colour to the young adult female recently taken in our harbour.

The three skulls alluded to are those of a very aged male from Shoalhaven, an adult female from Port Jackson, and an adult animal from New Zealand. These present the elongate form of the face, the mandible, *"strong with an acute angle behind,"* and the marked tricuspid form of the molars; but they differ, irrespective of size, in the following particulars:—

The occipital and sagittal crests of the aged male were comparatively greatly developed; the foramen magnum was actually smaller than in the other specimens, the aperture having become lessened apparently by the ossification of its upper portion; and the molars were set rather widely apart.

The New Zealand and Port Jackson specimens, the former especially,[1] had the two middle lower incisors and their sockets considerably within the outer ones; so that with the jaws closed they were completely over-lapped by the upper cutting teeth (the Shoalhaven animal did not possess this peculiarity); their molars also, were close together, almost crowded.

The old male measures twelve feet in length, and the skin presents a glossy, flecked appearance, of which the prevailing colours are light silvery grey, and very pale yellowish-white, arranged into numerous largish, longitudinal patches, occasionally brought into greater relief by a black-grey shading along their edges; the upper part of the body being darker, and the abdominal portions lighter than the sides.

The young adult female is in length seven feet two inches and a quarter, the upper half is darkish grey, becoming almost black along the dorsal line, and intermixed throughout with numerous narrow markings of darker hue, and of dull yellowish-white; under part, nearly unspotted, is of a dull dirty yellowish-white. These two colours do not blend at their junction, but, remaining distinct, pass from the tail, somewhat over the flippers, and immediately under the eyes and nostrils, to the end of the muzzle.

---

[1] Dr. Knox, of New Zealand, calls attention to this peculiarity.

The flippers, mostly the hinder, are anteriorly bordered by irregular black-grey markings, disposed transversely.

The Newcastle specimen, according to Mr. Angas, is seven feet ten inches in length, and in colouring faintly resembles the foregoing, but the neck and sides contain many distinct oblong black spots.

To these observations I may add that the specimen taken on the beach at Double Bay, within the harbour of Port Jackson, and which was kept alive for several days, on the grounds of the Museum, had the muzzle lengthened; the neck comparatively thin and long; and the girth of the body largest at the fore-flippers, which were placed somewhat in advance of half the entire length of the stretched out animal.

This species of Seal is by no means an infrequent visitor of the coasts of New South Wales and of New Zealand, but evidently only as a wanderer driven by untoward circumstances far away from home. No important details are recorded of its habits, further than that it is a resident of the colder portions of the Southern Seas, and that its capture is not carried on as an object of commercial enterprise.

Dr. George Bennett and Dr. Knox, however, give a few interesting particulars of the diet of this creature, when forced as an outcast to seek an existence on foreign feeding-grounds. The former, in his " Gatherings of a Naturalist in Australasia," p. 167, says, in allusion to the large male before described, that " it was killed in the fresh water of Shoalhaven River, in August, 1859, several miles above the influence of the salt water, and when opened had an entire water-mole in its stomach, minus the head." The latter observes of a New Zealand specimen, that "the stomach contained numerous fish-bones, a few feathers (gulls'), and some considerable portions of a pale-green broad-leaved marine fucus."

STENORHYNCHUS WEDDELLII, *Lesson.* False Sea Leopard of Gray.

Synonyms—*Leptonyx Weddellii*, Gray. B.M.C., 1866, p 12.
        Leopard Seal (?) Jardine's Nat. Libr., p. 183, pl. 12.
        Small-nailed Seal (?)   „   „   p. 180, pl. 11.

I derive the following extracts of the description of this species (the only one of Dr. Gray's *genus leptonyx*) from the British Museum Catalogue; putting the more salient points of distinction between it and *Stenorhynchus leptonyx* into italics, for the guidance of the student. "*Lower jaw weak, with an obtuse angle behind; orbits very large*"; head flattened; muzzle broad, rather short, rounded; muzzle hairy between and to the edge of the nostrils; nostrils ovate; whiskers compressed, slightly waved; ears, no external conch; skull slightly depressed, expanded behind; nose rather short, broad, high above; orbits, *rather large*; the petrose portion of the temporal bone convex, hemispherical. Lower jaw *slender*, with a short symphysis in front, and *narrow, without any angle at the hinder part of the lower edge.* The skull of this *genus Leptonyx* resembles in many respects Cuvier's figure of a skull of *Phoca-bicolor;* but it differs from it in all the grinders being placed more longitudinally, and in the *lower jaw being slender, and without any angle on the hinder part of the lower edge.*"

" Colour fulvous, with the front of the back and a line down the back blackish-grey ; whiskers brown, tapering. Female and young blackish-grey above ; sides with a series of longitudinal yellowish spots."

It will be seen that the external appearance of this seal resembles in part one or the other of the specimens I have just previously described, but the skulls in the Museum here all possess the lower jaw *strong, with an acute angle behind,* and the orbits *moderate.* The animal, which presents the marked characteristic of a weak and slender mandible, without any angle behind, implying a change in habit and in the nature of the food, must necessarily indicate a distinct species.

Inhabits South Orkney, Antarctic Ocean.

STENORHYNCHUS CARCINOPHAGA,[1] Homb. and Jacq. Crab-eating Seal.

Synonyms—*Lobodon carcinophaga,* Gray, B.M.C., 1866, p. 10.
     Stenorhynchus serridens, Owen.
     Halichœrus antarcticus, Peale.

This is a well defined species, distinguished from others of the genus by many properties, but principally by the altered form of the dentition : the first, second, and third molars of each ramus of the upper jaw, and the first one on each side of the lower jaw, are single-rooted—all the rest have two roots. The molars are compressed, and with swollen roots, the middle one with a large lobe in front and three lobes behind.

The symphysis of the mandible is said to be very long. Inhabits the Antartic Ocean.

"The head, back, hind feet, and upper part of the tail, pale olive ; fore feet, side of the face, body, and tail beneath, yellowish-white ; the hinder part of the sides of the body, and the base of the hind fins, yellow spotted, spots unequal often confluent."

STENORHYNCHUS ROSSII, Gray. Ross's large-eyed Seal.

Synonym—*Ommatophoca Rossii,* Gray, B.M.C., 1866, p. 14.

Another distinct kind.

Teeth throughout unusually small : "molars compressed, with a sub-central, rather large, broad, slightly incurved lobe, having a very small lobe on the inner side of its front, and a larger conical one in the middle of its hinder edge. Head short, broad ; ears small, with no external conch ; muzzle very short, rounded ; skull depressed, expanded behind ; orbits very large ; nose very short, broad, truncated in front, high behind."

" In colour greenish-yellow, with close oblique yellow stripes on the side, and pale beneath." Gray.

---

[1] καρκίνος, a crab, and φαγω, I eat.

Order 8.   DEINOTHERIA.[1]

Deinotherium,[1] Toxodon.[2]

Teeth of two kinds only, the canines being absent; bones dense; occipital region depressed, sloping from the condyles upwards and forwards; nasal aperture large, placed high up the skull; nasal bones short and salient; occipital condyles in the same line of direction with the longitudinal axis of the skull.

Family I.   DEINOTHERIOIDÆ.

Genus DEINOTHERIUM,[1] Kaup.

A cranium nearly perfect, and of about four feet in length, was discovered near Eppelsheim, in 1836, by M. Klipstein, in a sandstone deposit of the Meiocene period,—a period, I may observe, prolific in yielding peculiarly interesting fossil remains of species either wholly extinct or entirely superseded by new types, or of those still extant, but which seem to have now first sprung into existence,—such, for example, are the Deinotherium, the Mastodon, the Zeuglodon, and the Deer tribe of the present day.

It is from the structure of this remarkable skull that the following characters, descriptive of the family and probable habits of the animals when living, are arrived at; but until other portions of the skeleton are exhumed, the external form and the exact position of the Deinotherioidæ must remain a matter of simple conjecture.

$$\text{Incisors } \tfrac{0?-0?}{1\,-1}, \text{ canines } \tfrac{0-0}{0-0}, \text{ molars } \tfrac{5-5}{5-5} = 22 ?$$

The extremity of the upper jaw being mutilated, the presence or absence of the superior incisive teeth cannot be defined. The inferior incisors, however, are well preserved and highly characteristic; they are two in number, in close contiguity with each other, very large, tusk-like in form, with the ivory disposed in concentric striæ and embedded in enormous sockets. These, the tusks and sockets, bend abruptly downwards, almost vertically, maintaining, however, a gentle backward curve throughout. In the male the tusks are said to be two feet long, while those of the females are only about half that length. The molars are of comparative moderate size, and have their upper surfaces divided by two transverse ridges, excepting the middle one of each ramus and the first of the lower jaw; the former possessing three, and the latter only one, of these transverse ridges.

[1] δεινός awful, and θηρίον beast.
[2] τόξον a bow, and ὀδούς a tooth.

Of the skull the texture of the bones is dense; the occipital portion much flattened, with the hinder part inclined from before backwards; the nasal aperture large, and placed high up; the nasal bones short and salient; and the occipital condyles in the same line of direction with the longitudinal axis of the skull.

Such features are in themselves highly expressive of an aquatic existence, but they bear an additional value, not only confirmatory of the mode of life, but as suggestive of the structure of the body, from the marked resemblance presented by these several distinctive qualities to the similar ones seen in the skulls of the strictly aquatic animals, the Manatee and the Dugong; setting aside from the comparison the huge tusks and the lengthened sockets, to which singularities, although perfectly unique in their entirety, the Dugong affords a faint approach in the downward curve of the deflected symphysis of the mandible.

The natural inference, therefore, to be drawn from the cranial lineaments, by the absence of the canines, and by the form of the molar teeth, exhibited by this relic, would be—that the living Deinotherium giganteum partook the herbivorous habits of, and greatly resembled in general form, the members of the existing Sirenoid family.

Should this surmise, which I believe originated with de Blainville, be correct, this huge animal would possess a large, full, fleshy, trunk-less muzzle, adapted for browsing in shallow waters over beds of fluviatile or marine vegetation; nostrils advantageously placed at the end of the muzzle; and pinnated limbs, principally or wholly suited for progression in water.

The large massive skull, and the great weight of the incisors protruding from its extremity, are commonly urged as qualities materially affecting the probable terrestrial existence of the owner; but I think that such an argument is of but little value to come to any correct conclusion as to the economy of this extinct animal; for it is obvious that a frame-work fully equal to its requirements, and yet of no unusually stupendous dimensions, would render these apparent obstacles no more "cumbersome" or "inconvenient" to the quadruped on dry land, than the huge and weighty tusks[1] of the extinct Mammoth, which are supported with ease, or the vast expanding horns[2] of the fossil Elk of Ireland, borne with such graceful dignity.

Neither can I contemplate with the least satisfaction the form of the gigantic Deinotherium, as originally restored by M. Kaup; that is, an animal bearing an external resemblance to the tapir-like great Palæotherium, but with the lengthened proboscis of an elephant, the limbs of a rhinoceros, and feet terminated by the long hoof-like claws of a pangolin: truly a hetereogeneous compound, at variance with the significant characters displayed by the skull, and with the harmonious

[1] Each tusk 9 feet long, and weight of both 360 lbs.
[2] From tip to tip 12 feet.

organizations of animal bodies, where, for example, "the feet accord with the characters announced by the teeth; the teeth are in harmony with those indicated previously by the feet."

The immensity of organic fossil deposits, mostly fragmentary, many of wondrous shapes and generally of unallied kinds, promiscuously mingled, presents to the comparative anatomist a vast and too frequently a seductive field for imaginative speculation; and the fertile and heated brain, armed with such materials, is led to fabricate monsters as anomalous in their structural characters as those of heathen creations, or of the Middle Ages, and accepted as truthful by the credulous with an implicit belief.

So the pictorial illustration, acknowledged by the author himself to have been founded on error, still remains in works on Natural History, the stereotyped form of the Deinotherium giganteum.

Computing from this standard, the length of the animal is estimated at about 18 feet; but this magnitude would be greatly exceeded should it hereafter be ascertained that this singular being was truly aquatic in its habits.

Dr. Buckland suggests that the large incisive tusks served probably for tearing up and raking together the strong-rooted plants which grew in fresh-water rivers and lakes, and which probably constituted the diet of this pachyderm; for mooring purposes during repose; for dragging the immense carcass along the bed of the river or up its banks; and for weapons of offence and defence: in short, precisely similar in their uses to the effective upper canines of the Walrus.

Other fragmentary relics of the genus have been discovered in various parts of Europe and Asia, but their specific determination is still involved in considerable obscurity. The following list of species is about the best I can offer, although by no means so perfect as could be desired.

1. DEINOTHERIUM GIGANTEUM, Kaup.

    Syn.—*Tapir gigantesque*, Cuvier.
        Deinotherium maximum, Kaup.
          „   *medium*, Kaup.

2. DEINOTHERIUM CUVIERI, Kaup.

    Syn.—*Deinotherium bavaricum*, de Meyer.
          „   *secundarius*, Kaup.
          „   *Konigii*, Kaup.

3. DEINOTHERIUM MINUTUM, de Meyer.

4. DEINOTHERIUM PROAVUM, Eichwald.

    Syn.—*Tapirus proavus*, Eichwald.
        Mastodon podolicus, Eichwald.

5. DEINOTHERIUM INDICUM, Cantley and Falconer.

Family II. TOXODONTIDÆ.[1]

Genus TOXODON,[1] Owen.

The characters which distinguish this genus of Professor Owen have been derived by that naturalist from an imperfect skull, a few fragments of the lower jaw, and some teeth, discovered by Mr. Darwin, on the banks of the Sarandis, a small stream near Rio Negro, in South America. These distinctive qualities are principally as follows :—

$$\text{Incisors } \tfrac{2\text{-}2}{3\text{-}3}, \text{ canine } \tfrac{0\text{-}0}{0\text{-}0}, \text{ molars } \tfrac{7\text{-}7}{7\text{-}7} = 38.$$

The incisors have cutting edges, and are rootless, but supplied with persistent pulps ; of the upper ones, the two central are very small ; the two external, very large and curved ; the lower incisors have, on the contrary, the two middle large, with the others gradually diminishing in size. The molars, separated from the incisors by a wide interval, are rootless, curved (whence the generic name), and with an irregular central pillar of ivory, incased in a layer of enamel, which wearing unequally, give their surfaces an increased power of mastication.

The skull is massive and elongated ; the occipital region much depressed, and sloping downwards towards the condyles ; the occipital condyles in the same line of direction with the longitudinal axis of the skull ; the nasal aperture large, and placed high up ; the nasal bones short and salient ; and the cheek bones of great size and strength.

This animal, therefore—for there is only one species, the Toxodon platensis, *Owen,* whose remains are sufficiently known to represent the genus—assimimilates in many points to the animals of various other but distinct groups which exist at the present time.

To speculate even in a summary manner upon these counterpart characters is instructive, and may possibly intimate, by accepting the preponderating evidences, so adduced, the true nature of the form and habits of a singular animal, known but by a few imperfect relics.

It is said to resemble some of the extinct gigantic sloth-like quadrupeds, by the rootless and pulp-bearing molars, and by their massive construction ; but the presence of ten distinct incisors alone forbids the idea that any further affinity existed between it and the leaf-eating edentates, sufficient to justify the presumption that the limbs were furnished at their extremities with long subungulated claws.

In a somewhat greater degree it approaches the rodentia,—the form and composition of the cutting teeth, continually nourished by a pulpy substance, and the absence of canines, supplying the resemblance ; but the increased number of incisors, in direct variance with the typical character of the rodent, and the structural dissimilarity of the skull, lead to no inference that the feet were unguiculated to a similar extent, or that the general form was that of any one of the " gnawers."

---

[1] τόξον a bow, and ὀδούς a tooth.

Again, the number and peculiar disposition of the incisors, and the number and heavy make of the molars, point to a still nearer alliance with the Rhinoceros, and possibly with the water-loving anoplotherioids, whose canines are wanting, or are small and indistinct, and whose toes are protected by hoofs.

But the flattened crown of the head; the position of the breathing aperture, and that of the articulating process of the skull to the neck vertebræ, tend strongly to the conviction that the Toxodon, although from the presence of large frontal sinuses, was probably not so strictly aquatic as the Deinotherium[1], was nevertheless highly so, and nearly related to it, and to the Sirenioids; if so, the continual recurrence to the waters of the deep for subsistance would necessitate, as in the Seals, the use of fin-shaped limbs.

The skull in question measures about 2 feet 4 inches in length by 1 foot 4 inches in breadth, and about equals that of the Hippopotamus.

Toxodon angustidens, Owen.

Professor Owen regards the relics found at Buenos Ayres as constituting a distinct species, the animal of which would be but little inferior in size to the preceding.

Toxodon paranensis, D'Orbigny.

Is too much involved in obscurity to be considered as a reliable species.

Order 9. SIRENIA,[2]

Manatee, Dugong, &c., &c.

Teeth, when present, of two kinds only, incisors and molars; body elongated, conical, whale-like, sparsely covered with hairs; neck somewhat flexible; fore limbs converted into flippers, in some slightly unguiculated; fingers with the normal number of joints (three) as in the clawed mammals; hind limbs wanting, the body being terminated by an expanded, cartilaginous, horizontal tail; muzzle obtuse, truncated, thickly bristled; front of both jaws and part of the palate covered with a hard, corneous plate, externally tuberculated in undulating rows, the substance being composed of short, vertically placed bristles, agglutinated together by a horny matter, and bearing a considerable analogy to whalebone; nostrils separate, valvular, opening at the extremity of the muzzle, and connected to the nasal aperture of the skull by lengthened cartilage, and never employed as blow-holes; ears without

---

[1] Owen, Zoology of the Voyage of the "Beagle."

[2] *Sirenia*, from a supposed resemblance of the anterior part of the body, when raised out of the water, to that of a siren, or mermaid.

external conches, and orifices extremely small; eyes small, provided with nictating membrane ; mammæ two, pectoral ; voice reduced to a feeble lowing ; no dorsal fin or protuberance.

Of the skeleton, the bones are of dense texture, like ivory, and not loaded with oil; nasal aperture expanded, placed high up on the cranium; cheek bones massive; occipital condyles terminal; cervical vertebræ separate ; costo-sternal ribs cartilaginous; sternum composed of one piece; pelvis small, or rudimentary ; caudal regions elongated, possessing true V-shaped bones beneath their anterior vertebræ.

In habits the existing Sirenoids are gregarious, monogamous, (?) sluggish, usually frequenting shallow waters, and vegetarians in their diet.

The extant forms of this order are included within three well-defined genera, of which the species of two of them reveal, in their cervical vertebræ, a marked numerical deviation from the ordinary mammalian type, the three-toed sloth furnishing the only other exception to the general rule. In the Sloth these joints amount to nine, while in the present animals they number only six.

Family MANATIDÆ.[1]

(a) Genera DENTATA.

Teeth various in number; incisors large, conical, or, very small, early deciduous; molars at their apices flattened, transversely tuberculated ; posterior ones double-fanged ; lips single; stomach sacculated ; intestinal canal of great length; surface of skin, smooth, oily ; the two cavities of the heart at their lower ends separated from one another, each portion terminating in a distinct point.

The construction of the bruising molar teeth, the thick hide, and the great length and complicated nature of the intestinal canal, adapted for the digestion of vegetable food, ally these animals to the ordinary pachyderms, and consequently many zoologists have associated them with that group. Other writers have been induced to consider the affinity to approach nearer to the seals, from the bluff form of the head, the apical termination of the nostrils, the nictating membrane of the eye ; the lengthened neck, the more perfectly formed hand, and the density of the bones of the skeleton. Again, the peculiar nature of the layer of blubber, which envelops the muscular flesh and is immediately connected by cellular tissue to the external oily, almost naked skin, the entire absence of hinder limbs, the horizontally depressed cartilaginous extremity, the structure of the skeleton in

[1] From *manatus*, provided with hands.

E

almost every essential element, and the strictly aquatic life, appeal too strongly to the sense to admit of any doubt of their alliance with the Cetacea.

But beyond these connecting links, this singular group evinces, by the nature of its dentition, by the elevated position of the nasal aperture on the skull, useless as blow-holes, by the pectoral mammæ, and by many other deviating characters, so decided and so intrinsically different to similar parts of either of the orders enumerated, that of necessity a separation is required and a distinct locality assigned it among the mammals.

Influenced by this necessity for distinctive position, and guided by the greater alliance, shown in the osseous structure and in the habits of the living animal, to the seal and the whale, than to any terrestrial pachyderm, I have ventured to suggest that the natural allocation for the Order Sirenia should be between those of Pinnipedia and Cetacea.

Genus MANATUS, Rondelet.

Incisors $\frac{1-1}{0-0}$, canines $\frac{0-0}{0-0}$, molars $\frac{9-9}{9-9}$, $= 38$.

Incisors very small, early deciduous; molars squarish, irregularly flat on their apices, transversely tuberculated; of these several of the front ones frequently drop out, so that in the adult animal the number of teeth occasionally amounts to twenty-four only; front limbs terminated by small claws; tail rounded at its extremity; cervical vertebræ six; portion of the beak, anterior to the eye-sockets, short, advanced directly forwards, with a very slight gradual downward bend.

MANATUS AMERICANUS, Desmarest. The Manatee.

Synonyms—*Trichechus manatus*. Linnæus.
Manatus Americanus, Desmarest.
Manatus latirostris, Harlan.
Manatee (*i.e.*, Fish-ox), Negroes of Jamaica.
Coju-mero (*i.e.*, Sea-cow), Guiana.

In external appearance the Manatee is oblong, the body tapering from the shoulders posteriorly; the head is short, comparatively small, terminating at the muzzle with a thick fleshy disc, in the upper portion of which the nostrils are placed; the lips are studded compactly with stiff bristles; the front limbs are well developed, and possess a comparative free motion,—one, indeed, intermediate between the Seal and the Whale; small, flattish nails protect the tips of the fingers; of the hinder limbs there is no trace; the tail is cartilaginous, horizontally flattened, and rounded at its extremity. The colour of the adult varies from grey-black to blue-black, lighter and brighter underneath; the length from six to ten feet, even to fifteen feet; and the weights to correspond to these dimensions range from eight hundred pounds to a ton.

This species was at one time very abundant, delighting in the shallow waters of quiet bays and creeks of tropical America, and luxuriating in the sub-aquatic herbage, but more particularly about the mouths of the Amazons and the Orinoco, frequently ascending for many miles, even to their tributaries and the fresh-water lakes, where the floating leaves of water-plants supply their wants; but as the flesh, the hide, and the oil are much esteemed, and the animals themselves readily captured with the harpoon, the race has been greatly reduced by the assiduous persecution of the natives.

The male and female are said to be mutually attached so fervently, as to kill one the other becomes an easy prey, refusing to leave the fatal spot, and to forsake its late partner.

MANATUS SENEGALENSIS, Desmarest. The Lamantin.

Synonyms—*Lamantin du Sénégal*, Daubenton.
Trichechus manatus Africanus, Oken.
Manatus Senegalensis, Gray. Seals and Whales, 1866, p. 360.

The Lamantin inhabits the estuaries of the Senegal and other rivers of the western coast of tropical Africa, and, although considered to be distinct from the Manatee, it corresponds greatly with it in its organism, and apparently in its economy, of which, however, we have no sufficiently precise details.

Genus HALICORE,[1] Illiger.

Incisors $\male$ $\frac{1\text{-}1}{0\text{-}0}$ $\female$ $\frac{0\text{-}0}{0\text{-}0}$, canines $\frac{0\text{-}0}{0\text{-}0}$, molars $\frac{3\text{-}3}{3\text{-}3}$ = 12, or 14.

The upper incisors of the male are large, tusk-like, with bevelled off cutting edges, and their roots provided with persistent pulp-cavities, similar to rodents, and they project beyond their sockets only one-eighth of their whole length. In the female these teeth, although well developed, lie concealed, and never penetrate the gum; the molars during life number from twenty to twenty-four, but the first ones shed before the last has cut the gum, and consequently the whole are never simultaneously in use; the front limbs exhibit no trace of nails; the tail fin at its extremity is lunate, forked; the cervical vertebrae are seven; and that portion of the beak beyond the eyes, and which receives into its cavities the large upper incisive tusks, is bent down abruptly, almost vertically, over a long deflected mandibular symphysis.

The animals, therefore, are easily recognized from those of the genus Manatus, by either one of the many following characters, viz.:—the dental formula generally, but more especially the upper incisive tusks; the large and long facial bones, singularly bent downwards; the deflected symphysis of the lower jaw; the normal number of the neck vertebrae; the clawless, pectoral fins, and the forked extremity of the tail.

---

[1] ἅλς, the sea, and κόρη, a nymph.

HALICORE DUGONG, Illiger. The Dugong.

Synonyms—*Trichechus Dugong*, Gmelin, Pucheran, &c.
Indian Walrus, Pennant.
Dugong, Raffles, Home, Knox, &c.
Halicore Dugong, Gray, S. & W., 1866, p. 361.
Sea-pig of Moreton Bay, Captain Sidney.
Yangan, or *Yung-un*, Natives of North Australia.

The Dugong may be considered to be a tropical animal, although it is frequently seen in the waters of Moreton Bay, which would place it slightly without the verge of this prescribed limit. Its natural home, however, is in that extensive area embraced within both tropics, from the eastern coasts of Africa to those of Queensland. This vast range includes the Mauritius, Ceylon, the Bay of Bengal, the islands of the Indian Archipelago, and the northern coasts of New Holland, from the Barrow Reefs on the west round to Moreton Bay on the east.

But in these localities, it is only the shallow waters of unruffled inlets and creeks, the sheltered mouths of rivers, the bays and the straits between proximate islands, that afford the necessary quiet, and the abundant submersed marine aliment essential for a permanent residence. In such resorts the dugongs were formerly exceedingly plentiful, herding together in large numbers, and peacefully feeding like so many sheep on the seaweed, at depths of from six to twenty feet. They were at such times so far from being shy that, when rising at intervals to breathe, or drowsily basking on the surface, they allowed themselves to be handled,[1] so that the smaller and fatter ones were selected for food, and then shot at the end of the musket, or "laid hold of and forced on shore."

The natives of Sumatra, according to Sir Stamford Raffles, assert that the dugongs are never found on land or in fresh water, and their presence in shallows of the sea is at night-time ascertained by the snuffling noise they make at the surface. The Arabs also state of the dugong of the Red Sea[2] that they have feeble voices.[3]

Respecting these faculties, I have made many enquiries from well-informed persons, and the replies obtained confirm the truthfulness of the foregoing observations, at least when applied to the Australian animal. Hence, I cannot but think that the voice of the dugong scarcely exceeds the feeble lowing of the whale, and is not deep and hoarse as that of the larger seals, and that the fleshy front limbs of inadequate strength, the entire absence of hinder ones, and great unwieldiness of frame, substantiate the fact that this animal has not the

---

[1] Leguat, Penny Cyclopædia—Whales; confirmed by Mr. Edward Hill, who for, some time studied the habits of the dugong in the living state in Northern Australia. See also, in page 79, Steller's description of the Rhytina of the Arctic Seas.

[2] "This is probably the same as the dugong from India and Australia."—Gray, S. & W., 1866, p. 365.

[3] Rüppell.

power at will to travel inland in order to browse upon terrestrial herbage;[1] but whenever found in that position, it has been driven there by tempest, or, as Leguat pithily remarks, "laid hold of and forced on shore."

Péron observes, "the sailors were alarmed by a terrible howling, which resembled the roaring of a bull, but much stronger, and seemed to come from the reeds." And Mr. Fraser, in Captain Sterling's Voyage, 1826, notices that while attending to the boat on the river he "distinctly heard the bellowing of some huge animal, similar to that of an ox, from an extensive marsh further up the river."[2]

So far there is nothing extraordinary in these narratives, for the Australian eared-seal, Zalophus lobatus, is plentiful along the coast of Western Australia, and its habits and voice accord with the description of these travellers; but the explanatory remark attached by Dr. Gray, viz., "the roars were doubtless from the dugong," appears to be singularly infelicitous.

The dugong is a large, ungainly looking creature, reaching in its adult state to about twenty feet. The colour, according to some, is on the upper portions of the body slaty black; to others, brownish-black, with a whitish breast and belly. The skin is very thick and smooth, having over it a few remote and scattered hairs. The head is small, the upper lips are very large and obliquely truncated, on which part it is tuberculated and bristled; the flippers are short, thick, and fleshy, and incapable of supporting on land the huge bulk of the animal.

The flesh is much esteemed as an article of food by the natives of the various countries whose shores the dugong frequents, and the King of Malay claims as a royalty all those that are taken in his dominion. Many Europeans also affirm that it is excellent, comparing it to veal, to beef, or to pork; the variation in taste attributed to age and condition. But I am bound to add, that an opposite estimate of the quality of the flesh, whether roasted, boiled, or stewed, is entertained by some, to whom it appeared like coarse, oily bacon, and with a toughness and elasticity of fibre which kept the teeth engaged in a kind of perpetual motion.

The oil, however, is acknowledged by common consent to be of the best quality, peculiarly clear, limpid, and free from disagreeable smell.

These animals were formerly captured by the use of the spear, but of late years harpoons, nets, and boats with organized crews, have been

---

[1] "This induces them" (herbivorous whales) "to leave the water frequently to come on and crawl and pasture on the shore."—*Cuvier's Animal Kingdom.* "They" (dugongs) "are also found, and called 'the seal,' on the shore and in the salt-water inlet of the Concan, where they feed on the vegetable matter found on the rocks, and bask and sleep in the morning sun."—Gray, S. & W., 1866, p. 363. "Browsing on fuci, water-plants, or the grass of the shore."—Owen, Anatomy and Physiology of Vertebrates, 1866, vol. ii, p. 281. "Le Cétacé est mammifère qui vit dans l'eau, non comme le Phoque ou le Sirénien, qui prend librement ses ébats sur le rivage de la mer." Van Beneden & Gervais, Ost. des Cétacés, p. 1, now publishing.

[2] Gray, S. & W., 1866, p. 364.

employed in this country to meet the demand for the oil, which being carefully prepared, is said to possess all those remedial properties for which the cod-liver oil has become so noted.

Dr. Hobbs, of Brisbane, has the merit of first introducing this valuable commodity to notice, and it is much to be regretted that his enterprise has been cramped by the difficulty of procuring the Dugong in sufficient numbers, now that it has the experience of the ways of man.

Sir James Emerson Tennent, in his "Ceylon," states that the dugong there, while nursing, carries her offspring under one of her flippers, where the teat is situated, in such a position that the head of the young creature and her own are maintained above the water.

The Malays make frequent allusion to this animal as an example of maternal affection : when they succeed in taking a young one, they feel themselves certain of the mother, for she follows it, and allows herself to be speared and taken almost without resistance.

HALICORE TABERNACULI,[1] Rüppell. Abyssinian Dungog.

" Observed by Dr. Rüppell swimming among the coral banks on the coast of Abyssinia."

" The Arabs stated that they live in pairs,[2] or small families ; that they have feeble voices ; feed on algæ ; and that in February and March bloody battles take place between the males, which attain to eighteen feet."—*Penny Cyclopedia.*

Genus HALITHERIUM,[3] Kaup.

The more perfect remains which have been exhumed of this genus exhibit the osseous characters of the frame, similar to, and a cranium of very nearly the same form as the Dugong. The upper incisors assume the form of tusks, while the lower ones are very small ; the molars, however, approach nearer in structure to those of the Manatee, but their margins are deeply festooned ; of these, the upper have three tuberculated ridges,—the lower two ; all of the superior teeth are provided with three roots, and the inferior with two, the last of which is strongly fanged. The rib bones are solid, not porous or spongy.

These remains have been found principally in France, imbedded in the deposits of the upper Eocene groups, and those of the Meiocene period ; and one species was discovered at Piedmont, in the Pleiocene beds. From their marked analogy with the osseous characters of the Sirenoids,

---

[1] Of the tabernacle. So called by Dr. Rüppell from the notion that the Jews employed the skin in veiling the tabernacle.

[2] See habits of the Rhytina Stelleri, p. 56.

[3] ἅλς, the sea, and θηρίον, animal.

it is presumed that all the species of the genus Halitherium passed their existence in the shallows of the sea, or about the estuaries of rivers, and partook of the manners of the dugongs of the present day.

This genus was established by M. Kaup, to bring together many fragments incorrectly ascribed by Cuvier and others to different groups.

M. Pictet enumerates the following species:—

HALITHERIUM DUBIUM, Cuvier.

Hippopotamus dubius, Cuvier.
  „   *medius*,   „

HALITHERIUM GUETTARDI, de Blainville.

Vacha marina, Guettard.
Manatus Guettardi, de Blainville.

HALITHERIUM FOSSILE, Cuvier.

Phoca fossilis, Cuvier.
Trichechus fossilis, Cuvier.
Trichechus molassicus, Jaeger.
Manatus Cuvieri, Laurillard.
Manatus Cordieri, de Christol.

HALITHERIUM BEAUMONTI, de Christol.

HALITHERIUM STUDERI, de Meyer.

Metaxitherium Studeri, de Meyer.

HALITHERIUM SERRESSII, Gervais.

Cheirotherium, of Bruno.
Manatee, Dugong, Hippopotamus of various French authors.

Genus TRACHYTHERIUM,[1] Gervais.

M. Gervais founded this genus solely upon one tooth, which was obtained from the marine calcareous deposit of the Meiocene period at the Réole in France; consequently no information can be supplied beyond that it was the last molar of the lower jaw, and that it had on the crown three transverse tubercular ridges, and not two as in the corresponding tooth of the preceding genus, but otherwise resembling it.

Trachytherium Raulinii, Gervais, is necessarily the only species known.

(*b*) *Genus* EDENTATUM.

Genus RHYTINA,[2] Illiger.

Without teeth of any kind, the corneous lamellæ, before described, but on a larger scale, supplying their place; lips double, the outer

---

[1] τραχύς, rough, and θηρίον, animal.
[2] ῥυτίς, wrinkled.

upper one bristly; flippers small, clawless; tail-fin forked; stomach simple; surface of skin rough, folded, presenting a very rugged appearance (hence the generic name); cervical vertebræ six.

RHYTINA STELLEBI, Illiger.   Steller's Manatee.

Synonyms—*Manate, seu Vacca Marina*, Steller.
Trichechus manatus, Müller.
Trichechus borealis, Gmelin, Oken.
Rytina gigas, Gray, S. and W. 1866, p. 365.

Head small, oblong, obtuse; body dark-coloured, almost hairless, protected by a rugged covering, like the bark of an old oak, of which the scarf-skin is composed of fibres or tubes of a similar substance to the hoofs of cattle, closely packed, and perpendicular to and implanted into the true skin; the hide is an inch thick, and so tough as scarcely to be cut with an axe; but when cut appears like ebony in the inside; the tail is black, ending in a stiff, crescent-shaped fin, fringed with long fibrous matter like whalebone.

These curious animals, but a little more than a century ago, frequented in large herds the shoal waters of the bays and estuaries of the rivers of Behring's Straits, and of Kamtschatka, but are now probably wholly extinct by the ruthless hand of the seamen who were in the habit of wintering in these seas.

During Behring's second expedition, in 1741, Steller, who accompanied him, was compelled by shipwreck to remain on Behring's Island for ten months, and he estimated the then existing numbers to be so large as sufficient to supply food for the whole population of Kamtschatka. Sauer, the companion of the same great navigator in his third voyage, from 1789 to 1793, states that not a single specimen of the kind could be seen, the last known individual having been killed about twenty years previous to their visit.

Steller, however, has fortunately left behind him a comprehensive and reliable account of the habits and appearance of this singular being, and zoologists are thus entirely indebted to him for all the records they possess of a race, either effectually driven away from its natural haunts to more secluded homes, or, in accordance with the general belief, now numbered among the things of the past.

He relates, in addition to the characteristics given before, that they were so tame as to suffer themselves to be handled; if roughly treated, they removed to the sea, but soon forgot their injuries and returned. Sometimes they appeared in families near one another, each of which consisted of a male and female, one half grown, and a cub; these families often unite and form vast droves.

Their conjugal affection is most striking; a male, after using all his endeavours to relieve his mate, which had been struck, followed her to the water's edge, whence no blows could force him to depart. As long as she continued in the water, he attended; and even three days after her death, he was observed to remain in expectation of her return. They are most voracious creatures, and feed with their head under water, quite inattentivo to the boats, or anything that passes around them; moving and swimming gently after one another with a great portion of their back out of the water. Every now and then they elevate the nose to take breath, and make a noise like the snorting of a horse. They were taken at Behring's Island by a great hook fastened to a long rope, which was taken into a boat, and rowed among the herd. When the animal was struck, the loose end of the rope was conveyed to land, whore it was seized by about thirty people, who with great difficulty drew it on shore.

In summer they are very fat; in winter quite lean. Steller also observes that this animal grows to the length of twenty-eight feet, and that the weight of a very large one was 8,000 pounds.

ORDER 10. ZEUGLODONTIA.[1]

Teeth of two kinds (?), incisors conical, sharp-pointed, single-rooted, placed somewhat remote from each other; molars compressed, apex obtuse, double-rooted, set rather closely together into deep sockets; face of tho skull much elongated, slender; nasal orifices normal, that is, opening at the extremity of the muzzle, as in terrestrial mammals; body elongate, whale-like; pectoral limbs small, fin-shaped; hinder probably deficient.

From the foregoing general definition of the structure of the animals of this extinct group, it is interesting to observe that while retaining their own individuality, how singularly these organic remains connect by marked coincident features the preceding to the following order.

In instituting a comparison between them, we remark that the normal position of the breathing apertures, and the structural form of the teeth, are characters peculiarly their own; the teeth, most probably of two kinds only, with the molars, flattened at their apices, and double-rooted, attach them to the Sirenoids; the greatly elongated beak, and pointed incisors, suggest their alliance with the Whales; while the pisci-formed body and pinnated pectoral limbs are properties in common to them all.

Genus ZEUGLODON,[1] Owen.

Synonyms—*Basilosaurus*, Harlan.
        Squalodon, Grateloup.
        Hydrarchus, Koch.
        Dorydon, Gibbes.

---

[1] ζεύγλη, a yoke, and ὀδούς, a tooth—like two teeth tied or yoked together.

Incisors $\frac{3\text{-}3}{3\text{-}3}$, canines $\frac{1\text{-}1}{1\text{-}1}$}? molars $\frac{6\text{-}6}{6\text{-}6}=36$.

Incisors conical, pointed, curved inwardly, single-rooted; the nature of the teeth immediately adjoining these is doubtfully expressed in the dental formula, because many writers regard them as canines, while others only as premolars in the upper, and incisors in the lower jaw, the former being furnished with two roots and the latter with one ; the posterior molars are greatly constricted on both sides along the whole length of the middle of each tooth between the roots, and appear as if two teeth were united together by a slender bond ; indeed the form when viewed from above resembles rudely a seaman's watch-glass , and each is provided with two large fangs, whose length within the sockets occasionally doubles the exposed portion of the tooth.

ZEUGLODON MACROSPONDYLUS,[1] Owen.

Synonyms—*Basilosaurus*,[2] Harlan.
        Squalodon Grateloupi, Gervais.
        Delphinoides Grateloupi, Pedroni.

Each enlarged part of the molar teeth has a central pulp cavity and concentric striæ of growth ; the cranium is very much elongated and contracted from behind the frontals ; occipital region rises by an abrupt slope, as in cetaceans ; inferior jaw, in shape resembles those of the toothed-whales ; dorsal vertebræ are elongated and cylindrical ; cervical vertebræ short.

The fossil remains of this singular being was first discovered by Dr. Harlan, in 1835, at Arkansas, Mississipi, and he described them in the Transactions of the American Philosophical Society, under the name of Basilosaurus, as belonging to a gigantic reptile. Four years subsequently these bones were forwarded to Professor Owen, who, by a microscopic examination of the teeth, pronounced the animal to be an aquatic mammalian, a carnivorous whale, but likewise closely related to the Manatee. Other relics having been found at Alabama in 1845, Dr. Günther and many other zoologists supported Dr. Harlan's views as to the reptilian character of the remains, while Burmeister and Müller retained the mammalian theory. Since this latter date, bones in considerable quantities, and more perfect in distinctive qualities, have been exhumed, and these appear to confirm in a satisfactory manner the research and sagacity of Professor Owen in originally arriving at a correct conclusion from the inspection of materials apparently so slight and defective.

---

[1] μακρός, long, and σπονδύλος, dorsal vertebræ.
[2] βασιλεύς, king, and σαύρος, a lizard.

The species of this genus are but imperfectly determined, and much confusion in regard to their number exists among the comparative anatomists of the present day. Professor Owen treats only of two species, the *Z. macrospondylus* and the *Z. cetoides*,[1] to which M. Pictet and several others add the *Z. hydrarchus*,[2] Carus, *Z. trachyspondylus*,[3] Müller, and the *Z. Gibbesii?* (Dorydon[4] of Gibbes.)

Of these, the *Z.* Cetoides appears to be the largest, attaining in length to about seventy feet, although M. Koch exhibited a skeleton of one which measured 114 feet long ; this specimen, however, is considered to be made up by connecting the vertebræ of two distinct species, the *Z.* macrospondylus and *Z.* trachyspondylus, and consequently the accuracy of the foregoing dimension is now generally ignored.

The *Z.* hydrarchus varies from the *Z.* macrospondylus by possessing in the upper jaw four conical double-rooted teeth between the incisors and molars instead of two—and the *Z.* trachyspondylus is likewise distinguished from it by the rough and shorter formed vertebræ of the spinal column.

Genus SQUALODON, Grateloup.

The very imperfect remains of only an upper jaw yet found leave the question still doubtful whether this extinct species ought to be considered as a reptile or a cetacean. M. Grateloup, the originator of the genus, places it among the former ; while M. Beneden, Pedroni, and Agassiz assert its affinity to the latter. The number of teeth shown by these fragments amounts to at least $\frac{10 \cdot 10}{10 \cdot 10}$, for the tip of the beak is mutilated ; and incisors, known but by the alveoli, canines and molars, are distinctly exhibited, not doubtfully, as in the Zeugiodon. The molars are similar in their crowns to those of the Zeuglodon, but some of them are three-rooted. Until determined by better specimens, the Squalodon will have to occupy its present position.

SQUALODON GRATELOUPII, Gervais.

Squalodon grateloupii, Gervais.

Delphinioides grateloupii, Pedroni.

Phocodon (?), Agassiz.

Teeth $\frac{10 \cdot 10}{10 \cdot 10}$ (?), large, compressed, double or treble rooted ; muzzle not well known, elongated.

Discovered in the tertiary deposits in France and Austria.

---

[1] κῆτος, whale—whale-like.
[2] ὕδρα, sea-serpent, and ἀρχος, original, primitive.
[3] τραχύς rough, and σπονδύλος, dorsal vertebræ.
[4] δόρυ, a spear, and ὀδούς, tooth.

ORDER 11. CETACEA[1].

The order Cetacea comprehends all those animals so familiarly known under the trivial names of Whale, Black-fish, Grampus, Dolphin, and Porpoise, and which from their fish-like form and aquatic mode of life are, even at the present day, considered by a majority of persons as true fish.

Their genuine position among the first-class of vertebrated animals is clearly established by the one incontrovertible fact, that the female, bearing mammæ, suckles her young.

Beyond this important condition, there are, however, so many other interesting points of departure between the whale and fish tribes, so thoughtlessly associated together, that I cannot refrain from offering a few pertinent comparisons for the consideration of the student.

The young of the Cetaceæ are born alive, and such also is the case with those of many kinds of fish; but the young of the whale tribe, during the entire period of their fœtal condition, draw their nourishment direct from the parent by the agency of the connecting placenta, and after birth are wholly dependent upon the milk drained from her breasts; not so with the offspring of ovo-viviparous fish, the blennies, silures, sharks, &c., whose embryo is attached by a pedicle to a yolk, and both enclosed within an egg, externally protected by a membranous envelope. This egg enlarges within the body of the female, until, by the gradual and entire absorption of the alimental yolk, the fœtus is fully developed, and the young having thus been completely hatched, breaks the egg and the membrane which retained it, and passively seeks an independent living, conforming precisely in all the elements of fœtal vitality and after-growth to the young of the eggs scattered in the water by oviparous fish.

Again, the blood of all Cetaceans is warm, and consequently they are compelled to breathe the atmospheric air by means of true lungs, placed within the cavity of the chest, and have to rise periodically to the surface of the water in order to respire; should any accident frustrate this indispensable requirement, they would literally be drowned.

They are also provided with necks, which contain the typical number (seven) of vertebræ, common, with three exceptions only[2], to the whole mammalian group, although these necks are extremely short and thick; their eyes are protected by lids, and many species even possess the organ for secreting tears: their ears exhibit external openings; their nostrils are placed on the crown of the head; their caudal fin invariably assumes the horizontal position; and their body is covered with a smooth, almost hairless, leathery hide.

[1] κῆτος, a whale.
[2] The common Sloth (Bradypus tridactylus): the Manatee (Manatus Australis), and the Steller's Manatee (Rhytina Stelleri.)

On the contrary, the fish have the blood, although red, quite cold; their respiratory organs, termed gills, perform their functions through the medium of water, occasioning no necessity ever to quit that element to expel or inhale the breath ; the neck is totally absent, the head being immediately united to the trunk; the ears exhibit no external openings, as they are inclosed on every side by the bones of the head; the nostrils are at the end of the muzzle ; the body is usually covered with scales; the eye is lidless, and devoid of any lachrymal organ; and the tail is always vertically disposed.

Thus the Whales, in all the essential elements of organization, with the exception of external form, bear no more characteristic affinities to fish than the winged bats do to their co-tenants of the air, the birds.

The Whales have their bodies long and conical, terminating in a powerful, cartilaginous tail, the enormous muscles of which constitute their principle organ of locomotion; their mammæ are two, one on each side of the groin; the skin which covers the body is smooth and unprotected by hair, but it is immediately succeeded by a thick layer of blubber, which serves to keep up the proper temperature of the body requisite for the circulation of the blood ; the nostrils, or spiracles, are enlarged into blowers, capable of throwing up jets of water, or spray, accompanied by a loudish noise; the eyes are small, and situated towards the back part of the head ; the external orifice of the ear is minute, and can be closed at will; the front limbs, although displaying within all the ordinary bones appertaining to the hand, arm, and shoulder, are externally fin-shaped, and destitute of claws or nails, being thus rendered wholly unfit for grasping; the bones are, in general, very spongy, and strongly impregnated with fat, and in some respects resemble those of birds ; and the voice in most of the species is reduced to a simple lowing. The various kinds produce one young, occasionally two, at a birth, and the cubs follow their dams as calves do cows.

The cetaceans are carnivorous in the general sense, devouring animals from the small shrimp to a large seal. They are all, with a few exceptions, essentially marine, ranging the wide expanse of the ocean, and capable of remaining submerged for a considerable time.

There exists a remarkable peculiarity in the animals of this order, and which is exhibited even in the fœtal condition; namely, that the two halves of the head rarely correspond with each other in symmetry ; and it is likewise asserted that in some instances the bones of one side weigh heavier than those of the other. The greatest symmetry exists in the whalebone whales, and exactly the opposite prevails in the Zyphiidæ. Another remarkable departure from the mammalian type is found in the increase of the joints of certain fingers of the hand.

With reference to the toothed whales, Mr. Owen remarks that "the teeth exhibit no conformity of shape in the whole series, nor are they subject to succession or displacement by a second or permanent set," presenting "in the dental system but little fixity of character, and its variations extend in some cases to anomalies," yet it appears to me that, in their number and disposition, they afford to a certain extent sufficiently trustworthy indications of affinity for the classification of the species into several natural groups or families, and which, aided by other characters, would come, without being too elaborate, very conveniently within the compass of this primary treatise.

Somewhat contrary to modern practice, I head the following tabular classification of the Order Cetacea with the Family Platanistidæ, principally in obedience to the dental formula, and partly because of the many connecting forms of organic structure, which ally the members of this family with those of the two preceding orders—Zeuglodontia and Sirenia, on the one hand, and with the true Delphinidæ, their natural successors in the Cetacean line of affinity, on the other.

The known individuals of this fresh-water group approach the extinct carnivorous whales by their elongated head, of which the mandible exhibits a lengthened symphysis ; by their teeth firmly set in both jaws, the anterior ones retaining their prehensile character, while the posterior have their summits worn down broadly to their bases, the latter occasionally exhibiting two short fangs, a dental feature very exceptional among other existing whales.

They resemble the Manatee and the Dugong in possessing a comparatively long neck, of which the vertebræ are all free ; in bearing the cartilaginous costo-sternal ribs; in the peculiar mode of attachment of the ribs to the dorsal vertebræ; and in the sternum being composed of one piece.

The Inia, moreover, presents further corroborative affinities to the Sirenia worthy of notice, namely, in the restricted number of the lumbar vertebræ and the singular form of the cervical vertebræ and of the sternum,—these important parts being in their structure very unlike to the corresponding ones of any other cetacean, yet strongly recalling to mind those borne by the Dugong and Manatee respectively.

With the Delphinidæ the alliance is equally well marked—the small head, proportionate to the bulk of the body; the narrow, prolonged beak, parallelled in a less exaggerated form by some individuals of the genus Steno ; the two jaws of nearly the same length and breadth, and both armed with numerous trenchant teeth, the form of which in the Inia greatly strengthens the affinity with several species of the true dolphins, by having the surface of their crowns distinguished by a marked rugosity,—a singularity not only indicated clearly in the teeth of some species of the genus Steno, but seen in a lesser degree in those of the young of Orca and Pseudorca.

And in their manner of living and seizing their prey, no one can dispute the near relationship which exists between the Marine and Fluviatile Dolphins.

It is to be understood that the number, position, and nature of the teeth are taken from the adult animal only.

ORDER CETACEA.

TEETH.	FAMILIES.	GENERA.
SUB-ORDER 1. ODONTOCETE. TOOTHED WHALES.		
(a.) Megazoophaga, or Rapacious Whales.		
A { Permanent, numerous, occupying nearly the whole length of both jaws.	1. Platanistidæ............	Platauista. Inia.
	2. Pontoporiadæ	Pontoporia.
	3. Champsodelphidæ	*Champsodelphis. Arionius.*
	4. Delphinidæ	Steno. Delphinus. Tursio. Lagenorhynchus. Orcaella. Orca. Pseudorca. *Stereodelphis.* Phocæna.
	5. Delphinapteridæ ...	Delphinapterus. Neomeris.
(b.) Teuthophaga, or Squid-eaters.		
B { Deciduous, numerous, in front part only of both jaws.	6. Belugidæ	Beluga.
	7. Globiocephallidæ ...	Globiocephalus. Sphærocephœlus.
C { Long produced spiral tusks in upper jaw of male only.	8. Monodontidæ	Monodon.
D { Deciduous, in lower jaw only.	9. Grampidæ	Grampus.
	10. Hyperoodontidæ ...	Hyperoodon. Ziphius. *Choneziphius.*
	11. Physeteridæ	Kogia. Physeter. *Balænodon.*
	12. Mesoplodontidæ ...	Mesoplodon. Dioplodon. Berardius.
SUB-ORDER 2. ANODONTOCETE. WHALEBONE WHALES.		
(c.) Microzoophaga, or Insect-eaters.		
E { None; rudimentary and absorbed.	13. Balænopteridæ......	Balænoptera. Physalus. Sibbaldius.
	14. Megapteridæ.........	Megaptera.
	15. Agaphelidæ	Agaphelus.
	16. Balænidæ	Balæna. Eubalæna.

Before entering upon the details of the respective groups, taken seriatim as tabulated, I may remark that, with three or four exceptions only, in all the known species of the toothed whales(Odontocete) those which possess dorsal fins or distinct dorsal protuberances, have the cervical vertebræ more or less anchylosed[1]; whereas in those kinds where the dorsal fin is absent, the neck vertebræ are free[2]. The reverse of this singular union of two apparently unconnected, but important parts of organic structure occurs in the whalebone whales (Anodontocete), the dorsal-finned species having the cervical vertebræ in most instances free, while those without the back fin, with the exception of the scrag whales, have the vertebræ of the neck anchylosed into one solid mass.

Thus, by the examination of the cervical vertebræ of any cetacean skeleton set up in a museum, the dorsal outline of the living animal can with considerable accuracy be traced.

---

[1] Cervical vertebræ of Delphinus tursio (Tursio truncatus), are said by Cuvier to be free; Pontoporia blainvillii, cervical vertebræ free—Burmeister.

[2] By the same author (Cuv.), the neck vertebræ of Delphinus (Delphinapterus) Peronii are anchylosed, and also that of Delphinapterus (Neomeris) phocænoides.

Sub-order I. ODONTOCETE.

Toothed Whales.

Teeth always developed, but very variable in number and position; palate without baleen; head of moderate or excessive size; upper surface of skull unsymmetrical; external respiratory organ always single; gullet large; rami of mandible nearly straight, invariably connected anteriorly by a true symphysis; sternum nearly always composed of several pieces and united to several pairs of ribs by costo-sternal ribs of unossified cartilage or of bone.

(*a*) MEGAZOOPHAGA. Rapacious Whales.

A. Teeth permanent, numerous, occupying nearly the whole length of both jaws.

Family I. PLATANISTIDÆ.[1]

Without dorsal fin, back slightly keeled; cervical vertebræ free; head small, convex; both jaws of nearly the same length and breadth; beak very long and narrow; mandibular symphysis very long, more than half the entire length of the rami, greatly resembling that of the sperm whale; sternum of one piece; costo-sternal ribs cartilaginous; pectoral fin broad, truncated, five-fingered.

Genus PLATANISTA,[1] Pliny.

Teeth, $\frac{28\text{-}28}{28\text{-}28}$, or, $\frac{30\text{-}30}{30\text{-}30}$ = 112, or, 120.

Rather stout, set well apart, subcylindrical, anterior ones longest, compressed, and slightly curved; posterior ones short, with occasionally two short fangs. The long and slender beak, slightly more than three-quarters of the entire length of the skull, is compressed at the sides, expanded and slightly curved upwards at the extremity, where it is larger than in the middle. The black, shining eyes are exceedingly minute, scarcely one-eighth of an inch in diameter, obviously better adapted for turbid than clear waters. In general, their movements appear to be sluggish, but when in pursuit of fish they become active and fleet. There are but two species known, viz.:—

PLATANISTA GANGETICA, Roxburg. Dolphin of the Ganges. The Susu.

Synonyms—*Delphinorhychus Gangeticus*, Lesson.
Platanista Gangeteca, Gray, B.M.C., 1866; Supp. 1871.
Soosoo of the Ganges. Jardine, Nat. Libr., vol. 7.

The Susu of the Ganges attains to about seven feet in length, and its general colour is of a shining pearly grey. These animals frequent in

---

[1] πλατανιστης, Pliny's name for a dolphin, considered by Cuvier to be probably this very species.

great numbers the slow-moving labyrinths of the rivers and creeks which intersect the delta of the Ganges, but they are also known to have ascended that river to more than a thousand miles above Calcutta. They, however, confine their limits of range to within the bounds of rivers, never venturing out into the open ocean.

PLATANISTA INDI, Blyth.   The Susu of the Indus.

Synonyms—*Platanista Gangetica* (minor), Owen.   Cat. Coll. Surg.
Platanista Indi, Blyth. Journ. Asiat. Soc., Bengal: Gray, B.M.C., 1866; Suppl., 1871, p. 62.

As the name implies, this dolphin inhabits the river Indus and its tributaries, and in colour, size, and habits, bears a great resemblance to the species first described; it differs, however, in possessing a larger and more robust skull, and in the teeth, although equal in number, being shorter, and more ground down by attrition.

Genus INIA[1], Gray.

Teeth, $\frac{26\text{-}26}{26\text{-}26}$ to $\frac{33\text{-}33}{33\text{-}33}$ = 104 to 132.

Conical, permanent, firmly set, with compressed roots; anterior ones simple, sharp, slightly incurved; posterior with a broad, rounded tubercle towards the base of the crown; beak of the skull three-quarters of the entire length of the skull; pectoral fin large, ovate, obtusely pointed. The lower jaw, being terminated by a long cylindrical muzzle, affords, like the Platanista, an exact miniature resemblance to that of the Cachalot.

INIA GEOFFROYENSIS, de Blainville.   The Inia.

Synonyms—*Delphinus Geoffroyii,* Desm.; Mamm.
Delphinorhynchus frontatus. F. Cuvier.
Inia Boliviensis, D'Orbigny.   Voy. Amer. Merid.
Inia Geoffroyii, Gray, B.M.C., 1866. Suppl., 1871, p. 64.
Inia Geoffroyensis, Flower; Tran. Z. Soc., vol. 6, part 3.

This animal is at the present the only one of the genus, but from the great variation in the number of the teeth,[2] presented by several skulls in European museums, the probability arises that more than one kind will hereafter be distinctly determined.

The female Inia Geoffroyensis, when adult, measures about 7 feet; the male, it is said, arrives to a much larger size; the colour of both is of a pale blue on the upper portions of the body, and reddish underneath.

This dolphin is a native of South America, and, in groups of three or four, locates not only the remote tributaries of the Amazon, but, at vast distances from the sea, the elevated lakes of Peru, and thus may be considered as a true fresh-water cetacean.

---

[1] Inia, so named by the native Indians of Bolivia.
[2] On this point see Gray, Suppl., p. 64, and Flower T. Z. S., vol. 6, p. 87.

F

Family II. PONTOPORIADÆ.[1]

With dorsal fin; head small, convex, longly-beaked; jaws nearly of the same length and breadth, armed throughout with teeth; skull, long-beaked; beak slender, compressed, about three-fourths of the entire length of skull; mandibular symphysis very long, upwards of half of the entire length of ramus, greatly resembling those of Platanista and Inia; costo-sternal ribs ossified; cervical vertebræ free; teeth numerous, permanent, conical.

Genus PONTOPORIA,[1] Gray.

Beak high, compressed, slender smooth; pectoral fin short, truncated, five-fingered; dorsal fin short, triangular; mandible grooved on each side; symphysis frequently anchylosed by age; teeth with a swollen ring round the base.

PONTOPORIA BLAINVILLII, Treminville. The Pontoporia.

Synonyms—*Delphinus Blainvillii*, Tremin.
    Stenodelphis Blainvillii, Gervais.
    Pontoporia Blainvillii, Gray, S. and W. p. 231; Suppl.
    p. 96; Flower, Trans. Zool. Soc., vol. 6, p. 87.

Teeth $\frac{53\text{-}53}{53\text{-}53}$ to $\frac{57\text{-}57}{54\text{-}54} = 212$ to 222, small, in many respects resembling those of the Inia.

Inhab. coasts of Buenos Ayres—South Atlantic.

For many years a doubt has existed whether this species was fluviatile or marine, and whether it possessed a dorsal fin or was deprived of that appendage; for only a few remains, of uncertain origin, were known to the scientific world. Dr. Burmeister of Buenos Ayres has recently solved the problem, by the issue of a valuable monograph, whereby not only all doubts of its structure and habits are removed, but its position in the cetacean group is so far defined that it cannot be retained among the Platanistidæ, where some eminent writers had placed it, neither can it be classed with the Delphinidæ propriæ, as determined by other zoologists: the only alternative therefore appears to be to form an express family for its reception, and possibly for other allied kinds, recent or fossil, which some day may be brought to light.

It apparently constitutes an excellent connecting link between the fresh-water and marine dolphins, approaching the former by the prolonged beak, the greatly lengthened symphysis, and the free condition of the cervical vertebræ; and to the latter, by the dorsal fin and ossified cartilages of the ribs. I, therefore, assign to it the present position.

---

[1] πόντος, the sea, and πῶρις, a calf.

Family III. CHAMPSODELPHIDÆ.[1]

Beak of the skull much elongated; symphysis of lower jaw two-thirds the length of the entire ramus; teeth strong, permanent, numerously developed in both jaws.

Genus CHAMPSODELPHIS, Gervais.

Teeth with roots thicker than the crowns.

CHAMPSODELPHIS[1] MACROGENIUS,[2] Cuvier.

Delphinus macrogenius, Cuvier.
Champsodelphis macrogenius, Gervais, Pictet.

Is distinguished by the great length of the symphysis of the lower jaw, a peculiarity of structure only equalled by that of the sperm whale of our days; but the teeth being numerous and differently formed, prohibit the association of these two cetals. The dolphins, included in the family Platanistidæ, possess the mandibular symphysis rather more than half the length of the ramus, but Cuvier has shown that the construction of the bones more nearly allies this species to the Delphinus (Steno of Gray) rostratus, than to the Susu of the Ganges.

I may again be permitted to call attention to the fact that no existing species of the genus Steno, the most remarkable one among the Delphinidæ for exhibiting in excess this peculiarity, possesses a mandibular symphysis which ever attains to more than one-third the entire length of the ramus.

CHAMPSODELPHIS[1] BORDÆ, Gervais.

Is very similar to the foregoing, but found in a different locality.

The remains of these two species were discovered in France, embedded in the strata of the meiocene period; an age which also produced, among other mammals, those of the Deinotherium, the Zeuglodon (a carnivorous whale), and of the Halitherium (a Sirenian animal).

Genus ARIONIUS, H. de Meyer.

Teeth slightly bent; apices pointed; roots nearly circular.

ARIONIUS SERVATUS, H. de Meyer.

Synonym—*Delphinus molassicus*, Jaeger, Owen.

Found at Wurtemburg, in the molasse, or soft tertiary sandstone of the meiocene age.

---

[1] χάμψαι, a kind of crocodile, and δελφίς, dolphin—in allusion to the elongated form of the beak and formidable array of teeth.
[2] μακρός, long, and γένυς, the chin.

Family IV. DELPHINIDÆ.[1]

With dorsal fin; head small, convex, moderately beaked; symphysis of mandible moderate in length, never exceeding one-third of the entire length of the ramus; cervical vertebræ more or less anchylosed;[2] costo-sternal ribs ossified.

By a reference to the tabulated synopsis of the Cetaceæ, it will be seen that I have limited the members of this Family to only a few of the genera to be met with in Dr. Gray's last arrangement of 1871. Even as now restricted it is very fertile in species, and some or other of the individuals comprised within it may be met in almost every imaginable part of the ocean, north and south of the equator, enduring the extremes of heat and cold, disporting along the coasts, or within the shallower waters of the bays, ascending the innumerable creeks and inlets, or midway traversing the broad expanse of the sea, fleet and voracious, ever in a state of activity

> "Or dive below, or on the surface leap
> And spout the waves, and wanton in the deep."

The Delphinidæ, of which the smaller members bear to each other a most perplexing family likeness, are nevertheless made up of two tolerably well-defined groups, typically represented by the common dolphin and the common porpoise. The following general features will sufficiently distinguish the two. In the group of which the common dolphin is the representative, the head is more or less but decidedly beaked, the beak being parted from the forehead by a separating furrow; the lower jaw is usually of greater length than the upper one, and its symphysis moderately long. In the other section, typified by the porpoise, the head tapers uniformly towards the lips, scarcely, if at all, beaked, and without a divisional groove; the jaws are, in general, more of an equal length, and the symphysis of the lower one comparatively short.

With these few preliminary remarks, I proceed, in the concise terms to which I am limited, to the description of the genera and species of the Family, following in general the order of arrangement proposed by Dr. Gray; but deviating from it on such occasions, where I think the purpose I have in view is forwarded by a method of greater simplicity.

Genus STENO[3], Gray.

Head and forehead convex, moderately beaked; beak of the skull compressed, higher than broad, usually about $\frac{1}{11}$ of the entire length of the skull; symphysis of lower jaw elongate, from $\frac{1}{4}$ to $\frac{1}{3}$ length of the ramus; fore fins moderately long, triangular, obtusely pointed at the

---

[1] From *delphinus*, a dolphin.
[2] *Tursio truncatus* excepted ?
[3] στένος, narrow.

end. First finger short, cartilaginous: the second with six, the third with five, the fourth with two, and the fifth with one bony joint. The wrist bones all separated by broad cartilages. Shoulder-blade oblique, truncated at the posterior angle.—From Flower.

Teeth conical, large to small.

"This genus," says Dr. Gray, "is at once known from Lagenorhychus and Delphinus by the length, compression, and tapering form of the beak of the skull."

STENO FRONTATUS,[1] Cuvier. The Fronted Dolphin.

Synonyms—*Delphinus frontatus*, Cuvier, Owen.
Steno frontatus, Gray, B. M. C. and Suppl.

Teeth $\frac{24\text{-}24}{24\text{-}24}$ large, about two in an inch, often rather rugose.

Dr. Dickie describes the animal as having the skin rough, the back greyish black, and the belly dirty white, and the female being 9 feet long.

Inhab. Indian Ocean.

STENO COMPRESSUS,[2] Gray. The Narrow-beaked Dolphin.

Teeth $\frac{26\text{-}26}{26\text{-}26}$ large, about two in an inch.

"Beak of the skull compressed, attenuated in front."

Inhab. South Seas.

STENO ROSTRATUS,[3] Cuvier. The Beaked Dolphin.

Synonyms—*Delphinus rostratus*, Cuvier, Reg. Anim., p. 289.
Delphinus planiceps, Breda, 1829.
Delphinorhyncus Bredanensis, Jardine, N. L., vol. vii., p. 252.
Steno-rostratus, Gray, Suppl. B. M. C., 1871, p. 65.

Teeth $\frac{21\text{-}21}{21\text{-}21}$ or $\frac{23\text{-}23}{23\text{-}23}$, rather large, about two in an inch.

Animal black above, rosy-white beneath, having the lower part of the sides black-spotted.

Inhab. North Sea.

STENO SINENSIS,[4] Osbeck. The Chinese White Dolphin.

Synonyms—*Delphinus sinensis*, Osbeck, Desmarest, Flower, Tran. Zool.
Soc., vol. vii. p. 151.
Steno sinensis (Chinensis), Gray, Suppl. B. M. C., 1871,
p. 65.

Teeth $\frac{32\text{-}32}{32\text{-}32}$, many much worn down, about three in an inch.

The animal is milkish white, with pinkish fins and black eyes, and in length 7 feet 6 inches.

Inhab. China—harbour of Amoy, Canton, and Foochow Rivers.

---

[1] Fronted.
[2] Compressed.
[3] Beaked.
[4] *Sine*, China, adj. *sinensis*.

Steno Gadamu, Owen.   The Gadamu Dolphin.

Synonyms—*Delphinus Gadamu*, Owen, Trans. Zool. Soc., vol. vi., p. 17.
  Clymenia Gadamu, Gray, Suppl. B. M. C., 1871, p. 70.

Teeth $\frac{24-24}{24-24}$ to $\frac{27-27}{27-27}$, about three in an inch.
Colour of the animal, upper part plumbeous ashy-grey; underneath pinkish ashy-grey.   In length about 7 feet.
Inhab. Vizagapatam.

Steno maculiventer,[1] Owen.   The Spot-bellied Dolphin.

Synonym—*Delphinus maculiventer*, Owen, Trans. Zool. Soc., vol. vi.,
  p. 21.

Teeth $\frac{29-29}{28-28}$, about three in an inch.
Body above, a deep, shining, lead-coloured black; paler beneath, blotched irregularly over with ashy-grey.   Length about 7 feet; called by the native fishermen "Suvva."
Inhab. Vizagapatam.

Steno lentiginosus,[2] Owen.   The Freckled Dolphin.

Synonyms—*Delphinus lentiginosus*, Owen, Trans. Zool. Soc. vol. vi., p. 20.
  Steno lentiginosus, Gray, B. M. C., 1866, p. 394; Suppl.
  1871, p. 66.

Teeth $\frac{33-33}{32-32}$, about four in an inch.
Upper surface of the animal of a slate-colour, freckled over with lead-coloured streaks or spots; underneath similar but lighter.   Length nearly 8 feet; known by the native fishermen by the name of " Bolla Gidimi."
Inhab. Waltair, Vizagapatam.

Steno Malayanus, Lesson.   The Malay Dolphin.

Synonyms—*Delphinus Malayanus*, Lesson.   Voy. Coq.
  Delphinus plumbeus, Cuvier, Reg. Anim.   Jardine " Nat.
  Libr."
  Steno Malayanus, Gray.   B.M.C. 1866, p. 232.

Teeth $\frac{38-38}{38-38}$, about four in an inch.
Colour of the animal, greyish-lead throughout; length, about eight feet.
Inhab. Malabar Coast.

Steno Capensis, Gray.   The Cape Steno.

Teeth $\frac{37-37}{37-37}$, small, about five in an inch.
Mandible slender, attenuated; symphysis $\frac{1}{4}$ of ramus.   (See B.M.C. p. 394.)
Inhab. Cape of Good Hope.

---

[1] *Macula*, a spot, and *venter*, the belly.
[2] Freckled.

STENO ATTENUATUS, Gray. The slender-beaked Dolphin.
Teeth $\frac{40-40}{40-40}$, small, about five in an inch. (See B.M.C. p. 235, and Suppl. 1871, p. 66.)

STENO BREVIMANUS[1], Pucheran. The Singapore Dolphin.

Synonyms—*Delphinus brevimanus*, Pucheran. Voy. Dumont d'Urville.
Steno brevimanus, Gray. B.M.C., 1866, p. 236.
Teeth $\frac{36-36}{36-36}$, small, about five in an inch.
Inhab. Banda, Singapore.

STENO ROSEIVENTRIS[2] Pucheran. The Molucca Dolphin.

Synonyms—*Delphinus roseiventris*, Pucheran. Voy. Dumont d'Urville.
Steno roseiventris, Gray. B.M.C., 1866, p. 233.
Teeth $\frac{47-47}{44-44}$, small, about five in an inch.
" Greyish-black above, under half rosy-white ; orbit, streak from eye to the pectoral, and pectoral fin, dusky. Beak elongate, slender. Beak of skull very long, half as long again as the brain cavity." Gray.
Inhab. Molucca.

STENO FLUVIATILIS[3], Gervais. The Dolphin of the Upper Amazons.

Synonyms—*Delphinus fluviatilis*, Gervais and Delille.
Steno tucuxi, Gray. Suppl. B.M.C., 1871, p. 66.
Steno ? fluviatilis, *Steno ? pallidus*, Gray. Suppl. B.M.C., 1871, p. 66.
Teeth $\frac{29-29}{28-28}$ to $\frac{30-30}{30-30}$, small, about five in an inch.
" S. fluviatilis, above blackish, a broad band from the eye to the pectoral, and the pectoral fin black. Lower jaw and beneath rosy-white, the white bent up so as to form a broad white lobe behind the orbit over the pectoral." " S. tucuxi, dark blackish or fuscous ; it does not roll over like the Bouko (Inia Geoffroyensis), but comes to the surface to breathe : called by the natives *Tucuxi*." " S. pallidus, pale yellowish white above, beneath white." Both S. fluviatilis and S. pallidus " may be the same as S. tucuxi."
Inhab. " the upper parts of the Amazons River, 1,500 miles from the sea." See Gray. B.M.C. 1866, p. 236, 237. Suppl. 1871, p. 66.

STENO GUIANENSIS, Van Beneden. The Guiana Dolphin.

Synonyms—*Delphinus Guianensis*, Van Beneden.
Tursio ? Guianensis, Gray. B.M.C. 1866, p. 257.
Sotalia Guianensis, Gray. Suppl. 1871, p. 67.
Teeth $\frac{23-23}{23-23}$ to $\frac{30-30}{30-30}$, slender, conical, about five in an inch.
Inhab. British Guiana.

---

[1] *Brevis*, short, and *manus*, the hand.
[2] *Roseus*, colour of a rose, rosy, and *venter*, the belly.
[3] *Fluviatilis*, of or pertaining to a river.

Genus DELPHINUS.[1]

Head and forehead convex, longly-beaked; pectoral fin elongate, falcate, obtusely pointed; first, fourth, and fifth fingers short; second much the longest, with eight or nine joints; third about a fourth shorter than the second; beak of the skull elongate, commonly ⅘ of the entire length of the skull, depressed, broader than high; symphysis of mandible from moderate to short, usually between ⅓ and ⅕ length of the ramus; teeth conical, small, slender, set closely together.

(a) Palate behind deep channelled on each side.

* Teeth about five in an inch.

DELPHINUS LONGIROSTRIS, Dussumier. The Malabar Dolphin.

Delphinus longirostris, Dussum, Cuvier, Gray. S. and W. p. 241. Suppl. p. 68.

Teeth $\frac{55\text{-}55}{55\text{-}55}$ or $\frac{56\text{-}56}{56\text{-}56}$, about five in an inch.

Colour, black; length, nearly seven feet.

Inhab. Cape of Good Hope, Japan, Ceylon, and Malabar. Gray, B.M.C., p. 241. Suppl. p. 68.

DELPHINUS DELPHIS, Linnæus. The Common Dolphin.

Synonyms—*Delphinus delphis*, Linn., Hunter, Desmarest, Jardine, Cuvier, Gray, Bell, Nilsson, &c.

Teeth $\frac{42\text{-}42}{42\text{-}42}$ to $\frac{50\text{-}50}{53\text{-}52}$, about five in an inch.

Colour : black above, sides grey, and beneath white; length from 7 to 10 feet.

Inhab. the Northern and probably the Southern Seas.

This animal is gregarious in its habits, being very rarely seen alone, and is observed to be a constant attendant at such places where mackerel, pilchards, and other fish abound. It is swift and voracious, and so eager in the pursuit of its prey that it is frequently found entangled in the fishermen's nets.[2] The beak, which is characteristic of the group, is flattened and elongated, and from this peculiar form, it is called by the French, "Goose-bill and sea-goose." The English sailors, however, know this and similar animals by the name of "bottlenoses" or "flounder-heads," and not by that of *Dolphin*.[3]

Considerable doubt is entertained among naturalists, whether the Delphinus delphis of the European coast, and certain dolphins inhabiting the Southern Seas, are to be regarded as the same species. Possibly

---

[1] *Delphinus*, δελφίς, a dolphin.

[2] "In the month of September, 1845, eight or ten in a day were brought on shore in Mount's Bay, for many days in succession." Couch, Cornish Whales.

[3] The *dolphin* of sailors is a true fish, the Coryphæna hippuris, about five feet long, and celebrated for the eternal war it wages against flying-fish, and for the surprising changes of colour, when expiring.

not, but there is no appreciable difference in their general external appearance. Until this disputed point can be settled by the comparison of authentic specimens, I shall consider the two under the present denomination.

" These small whales (D. delphis) came around the bows of the ship in extensive shoals, and many were harpooned by our crew. We held them in much esteem for the table. When the external covering of lard or blubber is stripped off, the flesh beneath is found entirely free from fat or oil, and when cooked as steaks, bears a close resemblance to tender beef. It is certainly superior to the flesh of the turtle, cooked in the same form. The liver is also palatable and wholesome, and resembles the same part of a pig. In all the individuals we obtained, the contents of the stomach were either fish, cuttle-fish, or shrimps." [1]

DELPHINUS MOOREI, Gray. Moore's Dolphin.

Teeth $\frac{44\text{-}44}{43\text{-}43}$, five in an inch.

Upper surface of body, black; a black lunule-shaped streak from eye to eye, over the base of the beak; the sides and pectoral fins grey; chin and belly white; length, about 6 feet 4 inches.
Inhab. South Atlantic Ocean. Gray, B.M.C., p. 397. Suppl., p. 68.

DELPHINUS MARGINATUS,[2] Duvernoy. The Dieppe Dolphin.

Synonyms — *Delhinus marginatus*, Duvernoy, Desmarest, Gray, B.M.C., page 245.

Teeth $\frac{34\text{-}34}{43\text{-}42}$ to $\frac{49\text{-}49}{43\text{-}43}$? about five in an inch.

Upper part of the body black, paler on the head and sides; underneath, white; beak slender; teeth larger than common dolphin.
Inhab. Dieppe, Mediterranean ?

DELPHINUS MAJOR,[3] Gray. The Greater Dolphin.

Teeth $\frac{46\text{-}46}{47\text{-}47}$, nearly five in an inch.

Skull larger than that of the common dolphin. Gray, B.M.C., p. 396.

DELPHINUS JANIRA, Gray. The Janira.

Synonyms—*Delphinapterus Peronii*, Mus. Brist. Inst.
Delphinus Janira, Gray, B.M.C., p. 245, Suppl. p. 68.

Teeth $\frac{43\text{-}43}{42\text{-}42}$, about five in an inch.
Inhab: Newfoundland.

---

[1] Narrative of a Whaling Voyage round the Globe, 1833 to 1836, by Frederick Debell Bennett.
[2] *Marginatus*, broad-bordered.
[3] *Major*, greater.

DELPHINUS NOVÆ ZEALANDIÆ, Quoy et Gaimard. The New Zealand Dolphin.

Synonyms—*Delphinus Novæ Zealandiæ*, Quoy et Gaim., Voy. Astrol.
Delphinus Novæ Zealandiæ, Grey, B.M.C., p. 246.
Teeth $\frac{43-43}{43-43}$, five in an inch.

Above black-brown, like polished leather; the belly, and the edge of the upper jaw and the whole of the lower one dull white. A broad yellow band commences at the eye, narrows on the sides, and ends below the dorsal. The tail slate-colour. The pectorals are lead-white, like the middle of the dorsal, with black edges.

A black line from the upper part of the head, enlarging and inclosing the eye, which is bordered above and below with a white line. The eye large, black; the lower jaw with small rings of pores, and the body with small plaits of regularly twisted white striæ.

Length 5 ft. 10 in.

Inhab: New Zealand and Tasmania.

DELPHINUS ALBIMANUS,[1] Peale. The Chilian Dolphin.

Synonyms—*Delphinus albimanus*.[1] Peale.—Explo. Expedition, 1848.
Delphinus albimanus, Gray.—B.M.C., p. 247.
Teeth $\frac{43-43}{43-43}$, about five in an inch.

Colour blue-black; belly and pectoral fins white; sides pale tawny; eyes small, brown, surrounded with a black ring, which joins the black of the snout.

Length 6 ft. 6 in; weight estimated at 150 lbs.

Inhab: coasts of Chili.

This dolphin and that from New Zealand are presumed to be identical.

DELPHINUS PERNIGER,[2] Elliot. The Deep-black Dolphin.

Synonyms—*Delphinus perniger*, Elliot.—Journ. Asiat. Soc.
Delphinus perniger, Gray.—B.M.C., p. 249.
The teeth of this small dolphin are proportionally large.

Inhab: Bay of Bengal.

DELPHINUS FULVIFASCIATUS,[3] Pucheran. The Dusky-banded Dolphin.

Synonyms—*Delphinus fulvifasciatus*, Pucheran.—Voyage Dumont d'Urville.
Gray.—S. & W., p. 253, and
Teeth $\frac{47-47}{44-44}$ Supplement p. 68.

Blackish; side of back fulvous; throat and beneath white; beak, orbit, streak from angle of mouth to pectoral fin, and pectoral fin, blackish.

Inhab: coasts of Tasmania.

---

[1] *albus*, white, and *manus*, hand.
[2] *perniger*, very black.
[3] *fulvus*, deep-yellow, tawny, and *fasciatus*, swathed.

DELPHINUS POMEEGRA, Owen.   The Pomeegra Dolphin.

Synonyms—*Delphinus Pomeegra*, Owen.   Trans. Zool. Soc., vol. 6, p. 23.
Gray.   Suppl. S. and W., p. 69.

Teeth $\frac{43-43}{44-44}$, about five in an inch.

The upper portion of the body is of a shining deep leaden colour, almost black, becoming lighter on the abdominal parts.   It is described as a small cetacean.   The symphysis of mandible is less than one-sixth of length of ramus.

Inhab: coasts of Madras.

** Teeth between five and six in an inch.

DELPHINUS OBLIQUIDENS, Cope.

Synonyms—*Delphinus obliquidens.* Cope, Ac. Nat. Sc. Philad. 1869.
Gray, Suppl. S and W., p. 69.

Teeth ?

Inhab : North Pacific.

DELPHINUS SAO, Gray.   The Sao.

Teeth $\frac{63-62}{61-51}$ small, cylindrical, hooked.

Inhab : Madagascar.   From a skull in the Paris Museum.

DEPHENUS FRITHII, Blyth.   Frith's Dolphin.

Synonyms—*Delphinus Frithii*, Blyth. Asiat. Soc. Calcutta.
Gray, S. & W., p. 248.

Teeth $\frac{52-52}{50-50}$

Skull in the Museum of the Asiatic Society of Calcutta.

Inhab: "Procured during a voyage from England to India."

*** Teeth six in an inch.

DELPHINUS, WALKERI Gray.   Walker's Dolphin.

Teeth $\frac{47-47}{49-49}$, six in an inch.

"The pectoral fin, snout, the dorsal fin, a wavy streak from base of beak to eye, and upper surface of tail, black ; sides of the face and body to near the base of the tail, grey, with an elongated triangular patch beginning below the pectoral fin and extending near to the base of the tail, the broadest part over the vent.   Chin and beneath, as high as the base of the pectoral fin, and to the vent, white.   Length from end of snout to tip of tail 6 feet 7½ inches."   Gray, S. & W., p. 397.

Inhab. South Atlantic Ocean.

(*b*) Palate flat, not deeply channelled on each side.
* Teeth about five in an inch.

DELPHINUS STENORHYNCHUS, Gray.   The Steno-beaked Dolphin.

Synonyms—*Delphinus Stenorhynchus*, Gray.   S. & W., p. 396.
         Clymenia Stenorhyncha, Gray.   Supplement, p. 69.

Teeth $\frac{53\text{-}53}{53\text{-}53}$, five in an inch.
Hab: not known.

DELPHINUS FORSTERI?   Gray.   Forster's Dolphin.

Synonyms—*Delphinus Fosteri?*   Gray, S. & W., p. 248, Suppl. p. 69.

Teeth $\frac{45\text{-}45}{45\text{-}45}$ small, five in an inch.

Above, dark rust-coloured, beneath, dull dirty-white.

"Body straight, round, thickest behind; the pectoral fin tapering at both ends; head rounded, shelving in front, beaked: beak straight, pointed, cylindrical, depressed, attenuated, and blunt at the tip; upper jaw shorter, both blunt, toothed; eyes small, oblong, nearly in the middle of the side, near the gape of the mouth."   "Length 6 feet from nose to tail," (female).   Forster.

Inhab: "Pacific Ocean, between New Caledonia and Norfolk Island" Forster; Port Jackson, coast of New South Wales.

I have provisionally placed under this specific name a dolphin recently captured at Manly Beach, a short distance from the north entrance of Port Jackson.

This fine and perfect specimen Mr. Krefft, with his usual alacrity, at once secured for the Australian Museum, and, under his own super-intendence, when the animal was still alive, had photographs of its external appearance, and various admeasurements of the body, taken.

Our public institution consequently is now further enriched by another admirable skeleton of a cetacean, in addition to the several others lately acquired of these rare creatures, and also, with an un-usually correct stuffed representative of an extensive group, commonly so greatly caricatured in book illustrations and museum specimens.

Assisted by Mr. Krefft, I took the dimensions of many parts of the skull, and find that although these correspond closely in many particu-lars with the similar portions of the skull of Delphinus microps, now Stenorhynchus, marked *b*, detailed by Dr. Gray in his Catalogue S. and W. 1866, p. 240, yet the deviations betrayed by the comparison were sufficiently distinct to dispel any idea of their sameness.

I am, therefore, led to the belief when taking into consideration the size, external appearance, general colouring, with the exception of the small discal white spot on each fin, the geographic range of habitat, and common occurrence, that this dolphin is probably identical with the one described in 1774 by Dr. Forster, the companion of Captain Cook.

The dimensions alluded to are as follows, to which are added several other particulars :—

Teeth $\frac{45\text{-}45}{46\text{-}40}$, small, five in an inch.

Palate flat, without deep longitudinal channels on each side.

	Feet.	Inches.	Lines.
Entire length of skull	0	18	3
Length of beak, from tip to maxillary notch ...	0	11	8
Length of tooth-line, from tip to posterior edge of last tooth	0	9	11
Length of ramus of lower jaw	0	15	7
Length of symphysis of lower jaw (nearly one-seventh of ramus)	0	2	3
Width of beak at maxillary notch	0	3	10
Width of beak at its middle	0	2	2
Width of skull at the orbits	0	6	0
Length of animal, a female	6	8	6

Colour: above, dark-reddish brown, almost black along the back; beneath, dull dirty white.

DELPHINUS PUNCTATUS,[1] Gray. The Spotted Dolphin.

Synonym—*Delphinus punctatus*, Gray, S. and W., p. 398.

Teeth $\frac{40\text{-}40}{38\text{-}38}$ small, five in an inch.

"Upper portion black; sides with minute white spots; sides of the body above the base of the pectoral fin to the base of the tail blackish grey, which colour is obliquely extended as a lunate band from behind the vent to the back near the base of the tail." Gray, S. and W., p. 398.

Inhab: North Atlantic Ocean.

DELPHINUS EUPHROSYNE, Gray. The Euphrosyne.

Synonym—*Delphinus Euphrosyne*, Gray, S. and W., p. 251.
Delphinus Holböllii, Eschricht.
Clymenia Euphrosyne, Gray, Suppl., p. 70.

Teeth, $\frac{45\text{-}45}{42\text{-}42}$, about five in an inch.

Inhab: North Sea, Coast of England, South Atlantic.

DELPHINUS FRÆNATUS,[2] F. Cuvier. The Bridled Dolphin.

Synonym—*Delphinus frænatus*, F. Cuvier, Pucheran, &c.
Tursio frænatus, Gray. S. and W., p. 256.

Teeth $\frac{33\text{-}33}{34\text{-}34}$, about five in an inch.

Colour: above blackish; sides ashy; belly white; end of the tail black beneath; dark band from the angle of the mouth under the eye, whence the specific name; length 4 feet 6 inches. (This probably is a misprint.)

Inhab: Cape de Verd.

---

[1] Pierced with minute spots.
[2] Bridled, from *frænum*, a bridle.

DELPHINUS DUBIUS,[1] Cuvier.   The Variable Dolphin.

Synonym—*Delphinus dubius*, Cuvier, Reg. Anim., Gray, S. and W.,
    p. 253.

Teeth, $\frac{36 \cdot 36}{37 \cdot 37}$ about five in an inch.

Resembles the common dolphin in its colour.

Inhab: the Cape de Verd Islands.

DELPHINUS DORIS,[2] Gray.   The Doris.

Synonyms—*Tursio Doris*, Gray, S. and W., p. 255.
    Clymenia Doris, Gray, Suppl., p. 70.

Teeth $\frac{35 \cdot 35}{35 \cdot 35}$ or $\frac{36 \cdot 36}{36 \cdot 36}$, slender, five in an inch.

The skull only known.

Inhab: Cape of Good Hope.

DELPHINUS DORIDES,[3] Gray.

Synonyms—*Tursio Dorides* Gray, S. and W., p. 400.
    Clymenia Dorides, Gray, Suppl. p. 71.

Teeth $\frac{43 \cdot 43}{43 \cdot 43}$, small, slender, full five in an inch.

The skull only known.

Hab: unknown.

DELPHINUS EUTROPIA, Gray.   The Eutropia.

Synonyms—*Delphinus Eutropia*, Gray.   Zool. "Erebus" and "Terror."
    Tursio Eutropia, Gray, S. & W., p. 262.
    Eutropia Dichiei, Gray, Suppl., p. 75.

Teeth $\frac{34 \cdot 34}{33 \cdot 33}$, small, slender, about five in an inch.

From the skull only.

Inhab: South Pacific Ocean—Coasts of Chili.

DELPHINUS CAPENSIS, Cuvier.   The Cape Dolphin.

Synonyms—*Delphinus Capensis*, Cuv., Reg. Anim.
    Delphinus Heavisidii, Gray, 1828.
    Delphinus hastatus, F. Cuvier.
    Tursio Heavisidii, Gray, S. & W., p. 263.
    Eutropia Heavisidii, Gray, Suppl., p. 75.

Teeth $\frac{25 \cdot 25}{25 \cdot 25}$ to $\frac{28 \cdot 28}{28 \cdot 28}$, about five in an inch.

Black, with a white streak and two diverging lines beneath.

Inhab: Coasts of Cape of Good Hope.

---

[1] *dubius*, doubtful, variable.
[2] *Doris*, a nymph of the sea.
[3] Similar to, or resembling *Doris*.

DELPHINUS COMPRESSICAUDUS[1], Lesson. The Compressed-tailed Dolphin.

Synonyms—*Phocæna compressicauda*, Lesson, F. Cuvier.
Tursio compressicaudus, Gray, S. & W., p. 266.

Teeth $\frac{44 \cdot 44}{46 \cdot 46}$, small, about five in an inch.

Lead-coloured, belly whitish, base of tail compressed on each side.
Inhab : Western Coast, Africa. ?

DELPHINUS CHAMISSONIS, Weigm. The White-muzzled Dolphin.

Synonyms—*Delphinus chamissonis*, Weigm, Schreber.
Delphinus albirostratus, Peale, Expl. Exped.

Dark blue-grey, fins and back nearly black, a dark line connects the corner of the mouth with the pectoral fin ; front and sides dark grey, covered with small vermicular white spots ; end of snout white ; commissure of the lips pale yellow.
Inhab : Pacific Ocean.

DELPHINUS OBSCURUS[2], Gray. The Dusky Dolphin.

Synonyms—*Tursio obscurus*, Gray, S. & W, p. 264.
Clymenia obscura, Gray, Suppl., p. 71.
Phocæna australis, Peale, U. S. Expl. Exped.

Teeth $\frac{24 \cdot 24}{24 \cdot 24}$ to $\frac{26 \cdot 26}{26 \cdot 26}$, about five in an inch.

Colour, black above, with oblique diverging streaks on the sides ; underneath whitish.
Inhab : South Pacific.

** Teeth between five and six in an inch.

DELPHINUS CLYMENE,[3] Gray. The Clymene.

Synonyms—*Delphinus Clymene*, Gray, S. & W., p. 249.
Clymenia normalis, Gray, Suppl., p. 70.

Teeth $\frac{40 \cdot 40}{40 \cdot 40}$, small, nearly six in an inch.
Skull only known.
Inhab : ?

DELPHINUS LATERALIS[4], Peale. The Side-banded Dolphin.

Synonyms—*Delphinus lateralis*, Peale U. S. Expl. Exped. Gray, S. & W., p. 254.

Teeth $\frac{41 \cdot 41}{41 \cdot 41}$

A dark-coloured lateral line, edged by spots, separates the colours of the upper and under parts of the body, the latter being of a light

---

[1] *Compressus*, compressed, and *cauda*, the tail.
[2] Blackish, dusky.
[3] *Clymene.*
[4] *Lateralis*, belonging to the side.

purplish grey; there are also two other bands, paler in colour; the one branches from opposite the pectoral fin, and passes downwards and forwards; the other connects the eye with the pectoral fin.

Inhab: Pacific Ocean.

*** Teeth about six in an inch.

DELPHINUS MICROPS[1], Gray. The Small-headed Dolphin.

Synonyms—*Delphinus microps*, Gray, S. & W., p. 240.
Clymenia microps, Gray, Suppl., p. 69.

Teeth $\frac{48-48}{49-49}$, six in an inch.

Described from skulls only.

Inhab: Coasts of Brazils.

DELPHINUS STYX, Gray. The Styx.

Synonyms—*Delphinus Styx*, Gray, S. & W., p. 250.
Clymenia Styx, Gray, Suppl. S. & W., p. 70.

Teeth $\frac{42-42}{42-42}$, slender, six in an inch.

The skull very like to that of D. Euphrosyne, but the teeth more slender.

Inhab: W. Africa.

DELPHINUS TETHYOS, Gervais. The Tethyos.

Synonym—*Delphinus Tethyos*, Gray, S. & W., p. 251.

Teeth—?

Inhab: North Sea—South Atlantic.

DELPHINUS ALOPE, Gray. The Alope.

Synonyms—*Delphinus Alope*, Gray, S. & W., p. 252, 399.
Clymenia Alope, Gray, Suppl., p. 70.

Teeth $\frac{48-48}{46-46}$, very slender, six in an inch.

Inhab.: Cape Horn.

The skull only known.

The organic remains of several species, closely allied to this family and the preceding one, have frequently been discovered in the strata of the Meiocene period. Of these it is sufficient to notice that

The Delphinus pseudodelphis, Gervais, is so similar in the form of the skull and of the teeth to the Steno attenuatus that Dr. Gray suggests they may be of the same species.

The Delphinus dationum, Laurillard, and the D. vermontanus, Z. Thompson, approach in structure to the common dolphin; and

The Delphinus Renovi, Laurillard, greatly resembles the modern Delphinus longirostris.

---

[1] μικρός, small, and ὤψ, the face.

Genus TURSIO,[1] Gray.

Head and forehead convex, shortly beaked; beak of the skull short, stout, broad, depressed, slightly less than ½ but more than ¼ of the entire length of the skull; skull large, thick, heavy, with a high swollen brain cavity; symphysis of mandible short, from ⅓ to ½ of the length of mandibular ramus; teeth large; palate flat, not laterally channelled.

TURSIO TRUNCATUS,[2] Montagu. Bottle-nose Dolphin.

Synonyms—*Delphinus truncatus*, Montagu, Trans. Wern. Soc., London.
Delphinus tursio, O. Fabricius, Bonnaterre, Cuvier.
Tursio truncatus, Gray, S. & W. p. 258; Suppl., p. 74.

Teeth $\frac{21-21}{21-21}$ to $\frac{24-24}{24-24}$, three and a half to the inch; truncated when old.

Although this species is common in the European seas, yet there exists a considerable variance in the descriptions with regard to the colour of this animal, which probably may be accounted for by circumstances usually attendant upon age, sex, or season. It has been described as "black, whitish beneath"; "all blackish, the belly a little paler;" "uniform deep black," and "black, deeply tinged with purple, the sides dusky, belly greyish white." The length is commonly about 8 feet, but Cuvier states that individuals have been seen 15 feet long. In the old animal the skull becomes much thickened, the beak broad, flattened, and curved up at the tip in front.

Inhab: North Sea, the coasts of Great Britain and Ireland, the Mediterranean, &c.

TURSIO EREBENNUS,[3] Cope.

Teeth $\frac{23-23}{22-22}$, large, about two in an inch.
Inhab. Philadelphia.
See Cope, Proc. Ac. Nat. Sc. Philadel*.

TURSIO METIS, Gray. The Metis.

Synonym—*Tursio metis*, Gray, S. & W., p. 256; Suppl., p. 74.

Teeth $\frac{23-23}{22-22}$, conical, acute, curved.
Skull globular; beak thickish, conical, evenly tapering, upper part convex.
Inhab. West Africa.
Dr. Gray describes this species from a skull obtained during the voyage of the "Erebus" and "Terror."

---

[1] *Tursio*, a fish like a dolphin.
[2] *truncatus*, cut off.
[3] ἐρεβεννός, blackish, dusky.

G

Tursio Cymodoce,[1] Gray.   The Cymodece.

Teeth $\frac{22\text{-}22}{21\text{-}21}$, moderate, slightly incurved; rather more than three in an inch.
This skull is very like that of T. metis, but smaller.

Tursio abusalam, Rüppell.   The Abusalam.

Synonyms—*Delphinus abusalam*, Rüppell.
         Tursio abusalam, Gray.   S. & W., p. 261.   Suppl., p. 74.

Teeth $\frac{25\text{-}25}{25\text{-}25}$ to $\frac{30\text{-}30}{30\text{-}30}$.

Black, below white, with small dark spots.
Inhab: Red Sea, Cape of Good Hope.

Tursio Eurynome, Gray.   The Eurynome.

Teeth $\frac{25\text{-}25}{25\text{-}25}$.

Inhab: South Sea.   Bay of Bengal.
The skull is like that of Tursio truncatus, but the beak is longer, more slender, and more rounded.   The teeth also are smaller.

Tursio catalania, Gray.   The Cape Melville Dolphin.

Teeth $\frac{24\text{-}24}{23\text{-}23}$ to $\frac{27\text{-}27}{25\text{-}25}$, front lower teeth worn away, truncated.

"The adult, which was a female, measured $7\frac{1}{2}$ feet in length, and was of a very light lead-colour above and on the sides, gradually passing into the dirty leaden-white of the lower parts, which were covered, as also the flippers, with longitudinally elongated blotches of dark lead-colour. Another specimen, also a female, but much smaller, was exactly lead-colour, fading into whitish on the lower parts along the belly. The sides were marked with small oblong spots of the same colour as the back." *M'Gillivray.*
Inhab: North-west Coast of Australia.
This genus is represented in the collection of the Australian Museum by several fine skulls (one in particular from Wollongong), a mounted specimen, and an entire skeleton of an adult; the two latter being obtained from an animal captured in the waters of Port Jackson.
I am inclined to believe that the whole of these remains belong to individuals of the same species, and that that species, provided it be not the Tursio truncatus of the European Seas, is probably the one so well described in the foregoing extract, by our late fellow-citizen Mr. Macgillivray, and distinguished by Dr. Gray in his catalogue as Tursio catalania.
The wonderful similarity which exists in the anatomy of the dolphins of this group, and in the size and colour of the few actually examined while yet in the living state, renders it an almost hopeless task to

---

[1] Greek name of a woman.

recognize with any degree of certainty that specific individuality, so strongly insisted on in the able work so frequently cited. This difficulty is, moreover, increased by the knowledge that some of the so-called distinctive characters are possessed in common by all, while others are so feebly developed that they may fairly be attributable more to the varying conditions of life than to any essential structural quality.

I instance the worn-down or truncated teeth, a singularity, not the special property of any one, but equally belonging to all of the very aged ; and to the slight deviations in the form of the cranium, which, of themselves are so uncertain as to necessitate a continual revision of the systematic arrangement.

Beyond these methods but little remains to evidence any specific differences among this group, with the exception, perhaps, of "*locality*," the latter presenting but a false criterion by which any correct idea can be arrived at of aquatic animals, so wholly unrestrained by any barrier in their range of habitat.

Genus LAGENORHYNCHUS,[1] Gray.

Head convex, forehead low, gradually sloping into the beak ; beak short ; beak of the skull very short, from to slightly less than, half of the entire length of the skull, broad, depressed, narrowed in front, and bent up in front of the maxillary notch ; symphysis of mandible short, between ⅓ and ¼ of the length of the ramus ; teeth moderately large.

"This genus is easily known from Delphinus by the lowness of the forehead, the short and depressed form of the beak, the posterior position of the dorsal fin, the body being attenuated behind, and by the breadth and flat expanded form of the nose of the skull." Gray, S. & W. p. 268.

* Teeth three in an inch.

LAGENORHYNCHUS ALBIROSTRIS,[2] Gray. The White-beaked Bottle-nose.

Synonyms—*Delphinus albirostris*, Gray, 1846.

   Lagenorhynchus albirostris, Gray, S. & W., p. 272; Suppl., p. 79.

Teeth $\frac{25\cdot25}{24\cdot24}$, large, three in an inch.

"Upper part and sides very rich deep velvet-black ; external cuticle soft and silky, so thin and delicate as to be easily rubbed off ; the nose, a well-defined line above the upper jaw, and the whole of the under jaw and belly cream colour, varied with chalky white ; fins and tail black."

Inhab: North Sea. Faroe Islands. Yarmouth.

---

[1] Λάγηνος a cup, a flagon, and ῥύγχος, a beak, hence Bottle-nose.
[2] *albus*, white, and *rostrum* the beak.

** Teeth four in an inch.

LAGENORHYNCHUS ELECTRA, Gray.   The Electra.

Synonyms—*Lagenorhynchus electra*, Gray, S. & W., p. 268.
 Electra obtusa, Gray ; Suppl., p. 70.

Teeth $\frac{25\text{-}25}{24\text{-}24}$ moderate, four in an inch.
Species described from a purchased skull.
Inhab : (?)

LAGENORHYNCHUS ASIA, Gray.   The Asia.

Synonyms—*Lagenorhynchus Asia*, Gray, S. & W., p. 269.
 Electra Asia, Gray, Suppl., p. 76.

Teeth $\frac{24\text{-}24}{23\text{-}23}$, four in an inch.
Described from a skull only, which although the beak is rather more
attenuated and acute in front, Dr. Gray suggests that it may be only
a variety of the preceding species, the L. Electra.
Inhab : (?)

LAGENORHYNCHUS FUSIFORMIS,[1] Owen.   The Spindle-shaped Dolphin.

Synonyms—*Delphinus fusiformis*, [Owen.   Trans. Zool. Soc., vol. vi.,
 p. 22.
 Electra fusiformis, Gray.   Suppl., p. 76.

Teeth $\frac{22\text{-}22}{21\text{-}21}$, about four in an inch.
The upper portion of the body is of a light lead colour, fading into
light ashy-grey on the belly, and unspotted.   The dorsal and fore parts
of pectoral and caudal fins are much the darkest coloured.   The length
was about six feet.
Inhab : Waltair, Vizigapatam, India.

LAGENORHYNCHUS ACUTUS,[2] Gray.   Eschricht's Dolphin.

Synonyms—*Phocæna acutus*, Gray, 1828.
 Delphinus leucopleurus, Nilsson.
 Lagenorhynchus acutus, Gray, S. & W., p. 270.
 Electra acuta, Gray, Suppl., p. 76.

Teeth $\frac{32\text{-}32}{31\text{-}31}$, about four in an inch.
Above black, lower part of the beak and the body, shining white ;
a white band forms a line under the dorsal to the base of the tail ;
above yellow, beneath white.
The beak of the skull is more slender, and the teeth more numerous
than shown in those of any other species of this genus, yet this dolphin
is considered by some writers the same as the Lencopleurus, both
being about the same size, bearing a resemblance in the distribution of
the colouring, and inhabiting the same locality.

---

[1] *fusus*, a spindle, &c.
[2] *acutus*, pointed, in allusion to the slender form of the beak of the skull.

LAGENORHYNCHUS BREVICEPS, Pucheran. The Short-headed Lagenorhynchus.

Synonyms—*Delphinus breviceps*, Pucheran. Voy. Dumont d'Urville.
Lagenorhynchus breviceps, Gray, S. and W., p. 271.
Electra breviceps, Gray, Suppl., p. 76.

Teeth $\frac{31\cdot31}{29\cdot29}$ about four in an inch.
Blackish above, underneath white, pectoral fins dusky.
Inhab: Rio de la Plata.

*** Teeth five to six in an inch.

LAGENORHYNCHUS LEUCOPLEURUS, Rasch. The White-sided Bottle-nose.

Synonyms—*Delphinus tursio*, Knox. 1838.
Delphinus leucopleurus, Rasch. 1843.
Lagenorhynchus leucopleurus, Gray, S. and W., p. 273.
Leucopleurus arcticus, Gray, Suppl., p. 78.

Teeth $\frac{28\cdot28}{25\cdot25}$, nearly five to the inch.
Above bluish black, beneath white with a large oblique grey or white longitudinal streak on the hinder part of each side.
Of a female (the skeleton now in the Edinburgh Museum) of this species, captured at the Orkneys in May, 1835, Mr. Knox gives the following interesting particulars:—" It weighed 14 stone. Length from tip of beak to centre of tail, 77½ inches; weight of skeleton, 7½ lbs.; length of cranium, 15 inches; of spinal column, 55½ inches, equal to 70½ inches."
Inhab: North Sea, Orkney, Gulf of Christiana.

LAGENORHYNCHUS CLANCULUS, Gray. The Pacific Lagenorhynchus.

Synonyms—*Lagenorhynchus clanculus*, Gray, S. and W., p. 271.
Hector, Trans., N. Z. Institute, 1870, p. 27.
Electra clancula, Gray, Suppl., p. 77.

Teeth $\frac{33\cdot33}{32\cdot32}$, five in an inch.

LAGENORHYNCHUS CRUCIGERA, Gervais.

Synonyms—*Lagenorhynchus crucigera*, Gervais, Oib. Cet.
Electra crucigera, Gray, Suppl., p. 77.
Teeth (?)

LAGENORHYNCHUS THICOLEA, Gray. The Thicolea.

Synonyms—*Lagenorhychus Thicolea*, Gray, S. and W., p. 271.
Electra Thicolea, Gray, Suppl., p. 77.
Teeth $\frac{40\cdot40}{40\cdot40}$, slender, curved, elongate, six in an inch.
Described from the skull only.
Inhab. West Coast of North America.

LAGENORHYNCHUS CÆRULEO-ALBUS, Meyen. The Bluish-white Dolphin.

Synonyms—*Delphinus cæruleo-albus*, Meyen.
Lagenorhynchus cæruleo-albus, Gray, S. and W., p. 268.
Delphinus albirostratus, Peale, Expl. Exped.

Teeth $\frac{48-48}{50-50}$, about six in an inch.

Colour, "the back bluish, sides white, with oblique bluish streaks, belly white," in other specimens " dark blue-grey ; fins and back nearly black ; a dark line connects the corners of the mouth with the pectoral fins ; front and sides dark grey, covered with small vermicular white spots ; end of snout, white ; commissure of the lips pale yellow." Length from five and a half to six and a half feet. Inhab. Pacific Ocean. East Coast of South America.

LAGENORHYNCHUS ? INTERMEDIUS[1], Gray. The Small Killer.

Synonyms—*Delphinus intermedius*, Gray, Ann. Phil., 1827.
Orca intermedia, Gray, S. & W., p. 283.
Feresa intermedia, Gray, Supp., p. 78.

Teeth $\frac{11-11}{11-11}$, long, conical.
From a single skull in the British Museum.

Genus ORCAELLA,[2] Gray.

"Head blunt, rounded, very convex ; body moderate ; dorsal fin moderate ; pectoral fin broad. Skull : brain case, sub-globular ; beak very short, about ⅓ of the entire length of the skull, tapering, flat above ; palate flat in front ; rostral triangle very large, produced much in front of the maxillary notch ; lower jaw projects beyond the upper one ; symphysis short ; teeth small, slender, conical."

ORCAELLA BREVIROSTRIS,[3] Owen. The Short-beaked Killer.

Synonyms—*Phocæna brevirostris*, Owen, Trans. Zool. Soc., vol. vi, p. 24.
Orca brevirostris, Gray, S. & W., p. 285.
Orcaella brevirostris, Gray, Suppl., p. 80.

Teeth $\frac{14-14}{14-14}$, slender, conical.
Black ; body stout ; dorsal fin sub-centrical.
Inhab : estuaries of the Ganges ; Madras.
This species was first described in 1866, by Professor Owen, from the skull of a young animal, which was cast ashore in the harbour of Vizagapatam, on the east coast of India ; but since this period it has been observed by Dr. Anderson and Mr. Elliott. It is remarkable for

---

[1] Intermediate.
[2] Diminutive of *Orca*.
[3] *brevis*, short, and *rostrum*, beak.

the very short beak, for the slender, cone-like teeth, and in the living state for the round, almost globular top of the head, indicating its distinct nature and an approach towards the Globiocephalidæ.

ORCAELLA FLUMINALIS,[1] Anderson, M.S.   The Irawady Dolphin.

Synonyms—*Dolphin of the Irawady*, Anderson, P.Z.S., 1870, pp. 220, 544.

Orcaella fluminalis, Gray, Suppl., p. 80.

Body slender, dirty white.

Inhabits river Irawady, deep channels, from 300 to 1,000 miles from the sea.

Genus ORCA,[2] Rondelet.   The Killer.

Head rounded, scarcely beaked; dorsal fin high, falcate, central; pectoral fin broad, ovate; skull rounded; beak short, about one-half the entire length of skull; lower jaw strong, thick and solid in front, broad on the sides; symphysis moderate in length; teeth large, acute, flattened transversely, incurved at their tips.

On examining various skulls of the genera Orca and Pseudorca in our Museum, I was induced, from the variations presented, to analyze the tabulated admeasurements of others of various growths, recorded by Owen, Gray, Flower, Gervais, &c., as well as the excellent illustrations in Van Beneden's work on the Cetacea, now in the course of publication. From this research I arrived at the conviction that age and sex, assisted by occasional individual peculiarities, have produced many of those material deviations in the cranial structure which are so pointedly adduced as denoting distinct specific characters.

In illustration of this assertion,—select from either of these two genera (of course of the same species) the lower jaws of old and young animals, and it will be found by their comparison that in old age the length of the symphysis, the solidity of the adjoining parts, and the posterior span at the condyles, have respectively assumed proportions greatly in excess of those which might reasonably have been anticipated by a computation derived by rule of three from the condition of the similar parts of the young animal.

This additional massiveness of bone, and extra width of the grasping power of the mandible, beyond a proportionate increase in its length, would necessitate a corresponding change in the form of the cranium, sufficient to present a marked contrast between the skulls of the very aged and of the young adult.

---

[1] *fluminalis*, of or belonging to a river.
[2] *Orca*, the name given by Pliny to a large dolphin.

ORCA GLADIATOR,[1] Bonnaterre. The Killer.

Synonyms—*Delphinus orca*, Linnæus.
Delphinus gladiator, Bonnaterre.
Delphinus grampus, Owen.
Orca gladiator, Sundevall; Gray, S. & W., p. 279.
Ardluksoak is the name of the Greenlanders for the male,
and *Aidluik* for the female.

The males are much larger than the females.
Colour black above, shading into white on the abdomen, with usually
a more or less developed white patch above and somewhat behind the
eye.
The size of the adult males may be estimated at from 19 to 25 feet
in length, with a girth varying from 10 to 12 feet; but aged animals
have been captured which have measured 30 feet long. The body is
elongated and muscular, exhibiting a structure highly expressive of
speed and enormous strength.

NORTHERN VARIETIES.

ORCA STENORHYNCHA,[2] Gray, Suppl. S. & W., p. 90.

Teeth $\frac{11-11}{11-11}$; length of skull, 35 to 37 inches.
Colour of animal black; circumscribed spot behind the eye; spot on
belly; and under side of tail white; length 21 feet 3 inches.
Inhab: North Sea.

ORCA LATIROSTRIS,[3] Gray, Suppl. S. & W., p. 91.

Skull very similar to that of Orca capensis, but much smaller, and
distinguishable from the skull of the Orca stenorhyncha "by the
smaller size and broader, rounder nose."
Inhab: North Sea.

ORCA RECTIPINNA, Cope, Pro. Acad. Nat. Sc., Philad., 1869, p. 12.

Differs from the Orca stenorhyncha of Gray, and the Orca ater of
Cope, by having no white spot behind the eye.

ORCA ATER, Cope, Pro. Acad. Nat. Sc., Philad., 1869, p. 92.

Is known from being black above and below, but with a white spot
behind the eye.
Inhab: Oregon, Aleutian Islands.

---

[1] *gladiator*, a hector, a bully.
[2] From στενός, narrow, and ῥύγχος, beak.
[3] *latus*, wide, broad, and *rostrum*, beak.

ORCA ARCTICUS, Von Beneden and Gervais, Osteogr. Cet.
Skull, 24 inches in length.
Inhab: North Sea, Faroe Islands.

ORCA EUROPÆUS, Van Beneden and Gervais, Osteogr. Cet.
de l'Atlantic: teeth $\frac{12\text{-}12}{12\text{-}12}$; length of skull, 36 inches.
de Méditerranée: length of skull, 24 inches.

SOUTHERN VARIETIES.

ORCA CAPENSIS, Gray, S. & W., p. 283, Suppl., p. 90.

Synonyms—Delphinus globiceps, Owen.
Delphinus orca, Owen, Brit. Foss. Mamm.
Grampus, Bennett, Whaling Voyage.

Teeth $\frac{12\text{-}12}{12\text{-}12}$, Gray; $\frac{13\text{-}13}{13\text{-}13}$, Van Beneden.
Length of skull, 36½ to 37 inches, Gray; 42 inches, Van Beneden.
Colour similar to that of the Orca gladiator.
Dr. Gray observes, in page 285, " the examination of the skeleton, and
especially of the skull, shows that they (Orca gladiator of the British
coast, and Orca capensis) are quite distinct"; whereas Professor Owen,
in his elaborate analysis of the same animals, concludes his strictures
with the equally strong opinion: "The slight differences noticeable in
the skull chiefly depend on the muscular attachment, and are of a kind
to characterize varieties—not to establish specific distinctions." As I
entertain no belief in the theory of limited locality, particularly when
applied to such powerful beings, I adopt, without hesitation, the view
suggested by Prof. Owen, and consider the O. capensis as a variety only.

ORCA AFRICANA, Gray, Suppl. S. & W., p. 91.
Skull, 24 inches in length.
Inhab: Algoa Bay.

ORCA MAGELLANICA, Burmeister, Ann. & Mag., Nat. Hist. Gray,
Suppl. S. & W., p. 92.
" This species, according to the figure, is very like Orca latirostris."
Inhab: Patagonia.

ORCA TASMANICA, Gray, Suppl. S. & W., p. 92.
Skull, about 32 inches in length.
Inhab: Tasmania.

OPHYSIA (ORCA) PACIFICA, Gray, Suppl. S. & W., p. 93.

Synonyms—Delphinus orca, Eydoux.
Orca capensis, Gray, S. & W., p. 283.
Teeth ?
" Intermaxillaries very narrow, slightly dilated in front; brain-cavity
broad; occiput deeply concave; lower jaw very broad on the sides, very
thick and solid in front."

A very fine mandible in the Australian Museum belongs, I think, to this new genus, for it is unusally "very broad on the sides, very thick and solid in front"; to which I may add, and comparatively very wide apart from condyle to condyle. I can, however, trace in it no radical characters, beyond those I attribute to the natural results of age.

This lower jaw presents the following particulars :—Entire length of ramus, 33 in. 9 lines. Teeth $\frac{12\text{-}12}{12\ 12}$, very perfect ; anterior and posterior pairs small, respectively increasing in size towards the central portion of the tooth-line, where they become very large, conical, acute, and slightly incurved at their tips, and towards the base flattened transversely ; symphysis in length, 8 in.; height of ramus at coronoid process, 10 in. ; length of tooth-line from tip to posterior edge of last tooth, 16 in. ; posterior spread of jaw, meaured from outward edges of the condyles, 24 in. 4 lin.

The owner of this formidable mandible must have possessed dimensions in other portions of its structure equal to those entertained by the huge Orca capensis described by Van Beneden and Gervais, and greatly exceeded in magnitude, not only all of Dr. Gray's accredited species of the genus, but that of his Ophysia pacifica.

Where the sexes assume such widely different growths, why may not the larger skull in our museum, and the smaller one of the O. pacifica be fairly considered as sexual characteristics of the same species, of which the individuals have been so fortunate as to reach an unusually mature age.

Inhab: North and South Pacific Ocean.

Whatever differences of opinion (and they are great) that may exist among zoologists as to the number of species, all bestow upon the members of the family the qualities of vast strength, great ferocity, and an insatiable appetite. Pliny, more than 1,800 years ago, recorded the ferocious habits of this rapacious whale, which Linnæus[1], Otho Fabricius, Nilsson, and other naturalists, have fully confirmed.

From the more recent observations of reliable men, I select the following interesting accounts of the habits of the killers. Captain Holböll states that, "in the year 1827 I was myself an eye-witness of a great slaughter performed by these rapacious animals. A shoal of Belugas had been pursued by these blood-thirsty animals into a bay in the neighbourhood of Godhavn, and were there literally torn to pieces by them. Many more of the Belugas were killed than eaten, so that the Greenlanders, besides their own booty, got a good share of that of the killers. In the year 1830, a large Krepokak (Megaptera longimana) was overpowered by an Orca, in the neighbourhood of Narparsok, according to the statements of the Greenlanders, and torn to pieces after it was dead. Almost fifteen barrels of the blubber, floating about at the place where the struggle had taken place, fell to the share of the Greenlanders. It is principally the blubber that is the most coveted food of the killers,

---

[1] " Balænarum Phocarumque tyrannus, quas turmatim aggreditur."

not the tongue, as I have stated in several places. In this Krepokak especially, the tongue was found untouched, and was afterwards flensed by the Greenlanders." " They are able to swallow whole porpoises, as well as seals, even very large individuals, four at least immediately after one another (according to Nilsson's observation), and in the course of a few days as many as twenty-seven individuals; nor do they fear to attack and tear to pieces the very largest whalebone whales, in order to satisfy their hunger on the blubber."[1]

Mr. R. Brown, who, from his position in the Arctic Regions, and long experience among the northern whales, must be regarded as a good authority, also notices in the Proc. Zool. Socy., 1868, that "the Aidluik is only seen in the summer-time along the whole coast of Greenland. Wherever the white whale, the right whale, or the seals are found, there is also their ruthless enemy, the Killer.[2] The white whale and seals often run ashore in terror of this Cetacean, and 1 have seen seals spring out of the water when pursued by it." "Though subsisting chiefly on large fishes, they will not hesitate to attack the largest whalebone whales, and are able to swallow whole, large porpoises and seals." " I know of a case in which they attacked a white-painted herring-boat in the Western Islands, mistaking it for a Beluga."

The Southern Killers have similar habits; they associate in groups, and follow up their prey, the larger in the more open ocean, the smaller in bays and shallower waters.

Even to the most fastidious imagination, not hopelessly prepossessed, can there be presented any possible barrier which would prevent beings, endowed with such physical powers, emigrating, when overstocked, to other feeding grounds, and thus, in the course of generations, spreading out into families over the face of the ocean?

Genus PSEUDORCA,[3] Reinhardt.

The generic characters are similar to those of the genus Orca, with the exception that the dorsal fin is moderate, and the pectoral fin small, falcate, not of that great breadth so peculiar to the larger killers. The mandible is strong, but does not exhibit at its front the depth and solidity of that of the Orca. The array of teeth is even more formidable from their strength and solidity, although the animals carrying them are much smaller than the foregoing genus.

PSEUDORCA CRASSIDENS,[4] Owen. The Lincolnshire Killer.

Synonyms—*Phocæna crassidens*, Owen, Brit. Foss. Mamm.
      Pseudorca crassidens, Reinhardt, Gray, S. & W., p. 290,
        and Suppl., p. 80.

---

[1] Eschricht on the Northern species of Orca. Ray Society—1866, p. 168.
[2] "They (the Grampus of the South Seas) occur in herds, and their appearance is supposed to indicate the resorts of the Cachalots." Bennett, Whaling Voyage.
[3] False Orca.
[4] *crassus*, thick, and *dens*, tooth.

The English name is derived from the discovery, in 1840, of the fossil remains, in the fens of Lincolnshire, of a whale, of which the living representative was then unknown. The skull, shortly after the discovery, was fully described by Prof. Owen, in the "British Fossil Mammalia."

Twenty years subsequently, M. Reinhardt, a Danish naturalist, found a species of whale existing in the northern seas, which, by the careful examination of its skeleton, he ascertained to be identical with Owen's Pseud. crassidens.

"I therefore believe," observes M. Reinhardt, "that we must really acknowledge this Phocæna crassidens of Owen to be the dolphin stranded on our coasts; however strange it may seem that our first knowledge of a Cetacean, of which great shoals are still in our time roaming about in our Northern Sea, should have come to us through an individual which thousands of years ago found its resting place on a sea bottom, now forming part of the soil of England."

There were two of these whales stranded on the shore near Asnæs, on the Danish coast, and they measured 14 and 19 feet respectively.

Teeth $\frac{8\text{-}8}{10\text{-}10}$, large, conical; rather acute.

Colour—black, paler below.

Inhab: North Sea.

SOUTHERN VARIETY.

Pseudorca meridionalis,[1] Flower. The Tasmanian Killer.

Synonyms—*Pseudorca meridionalis*, Flower, Pro. Zool. Soc., 1864; Gray, S. & W., p. 291; Suppl., p. 79.

Black Grampus of whalers, Crowther.

Teeth $\frac{8\text{-}8}{10\text{-}10}$, large, conical, acute; smallest in front, largest in the central portion. Symphysis of mandible nearly ⅓ of the entire length of the ramus.

Colour, black, with the under portions whitish.

Males much larger than the females, and similar in size to the animals of the preceding species.

Inhab: South Sea: coasts of Tasmania.

"To find distinctive characters to separate the present species from O. crassidens is a matter of greater difficulty." "The beak is much more pointed at the extremity, and the premaxillaries are of different form." "I think that these are sufficient, together with the great improbability of the same species being found in such widely different regions, to justify my regarding the small grampus from Tasmania, however familiar to the inhabitants of that country, as a species new to zoological literature, and imposing upon it the name of Orca (Pseudorca) meridionalis."

---

[1] Southern.

Tho skulls in the Australian Museum of the Tasmanian killer vary somewhat in themselves, and neither of them exhibit the differentiating character of a more pointed beak than that possessed by the Pseud. crassidens, in so marked a manner as the figure in the Proc. Zool. Soc., London, of the skull in the Royal College of Surgeons. Indeed, one of them, by the more rounded form of the terminal portion of the beak, assumes in its cast of features a position intermediate between the animals described by Messrs. Reinhardt and Flower; thus lessening, by the gradual approach to cranial similitude, the slight specific distinctions presented by the examination of two skulls only by the latter zoologist.

Setting aside these trivial shades of facial disagreement, particularly in a class of animals notorious for cranial variations, and taking into consideration that in all the essential qualities of osseous structure, in the size, in the colour, and in the habits of the living animal, both so-called species intimately agree, I am irresistibly led to the conclusion that the northern and southern animals are identical, or at the most, but varieties caused possibly by some lengthened period of isolation from others of their kindred.

" The Tasmanian killer, black grampus, or peaked-nose blackfish, is usually met with in shoals ranging from fifty to one hundred each, and always in company with the smaller kinds of the delphinidæ and cow-fish" : I presume, on the principle " wheresoever the carcass is, there will the eagles be gathered together." " It is a peculiarly wary cetacean, and I never heard of any having been captured, with the exception of one fastened to at day-dawn in the North Pacific Ocean, by an American whaler, supposed at the time to be an ordinary school sperm whale." [1]

A good mounted skeleton of this animal in our Museum gives an entire length of 16 feet, that of the skull as somewhat exceeding 25 inches, and of the beak measured from a line drawn between the maxillary notches to the tip as 11½ inches.

Genus STEREODELPHIS,[2] Gervais.

Teeth rather large, with short, nearly hemispherical crowns.
Stereodelphis brevidens, Dubreuil et Gervais.
The fossil remains of this, the only species, were found in the soft tertiary sandstone at Hérault, one of the departments of France.

Genus PHOCÆNA.[3]

Teeth permanent, compressed, sharp-edged, rounded at their tips; head rounded, scarcely beaked, without any separating furrow; muzzle uniformly rounded to the extremity, where it gently curves upwards,

---

[1] The information respecting the habits of the Tasmanian Killer, contained in these two quotations, was kindly supplied by my friend W. L. Crowther, M.D., M.L.C. of Hobart Town.

[2] στερεός solid, and δελφίς, dolphin.

[3] φώκαινα, a porpoise.

forming in its union with the gums a thickened pad, resembling lips; when the mouth is closed, the upper lip overlaps the under one evenly all round; pectoral fin elongate, obtusely pointed; dorsal fin central, slightly spined or tuberculated anteriorly.

PHOCÆNA COMMUNIS,[1] Brookes.   The common Porpoise.[2]

Synonyms—*Delphinus phocæna,* Linn. ; Bonnat ; Desm. ; Cuvier ; Bell ; Turton ; Fleming ; Nilsson.

Phocæna communis, Brooks ; Gray, S. & W., p. 302, Suppl. 81.

Phocæna tuberculifera, Gray, S. and W., p. 304.

Teeth $\frac{26-26}{25-25}$, compressed, rather stout.

Colour of the upper portion of the body, a deep bluish black, fading away on the sides, and becoming silvery white on the abdomen.   In length it ranges from four to six and a half feet, being probably the smallest cetacean known.

Inhab: North Sea, mostly in shore, frequently ascending rivers as far as the waters continue salt.

The porpoise was well known to Pliny, who described its form accurately; and in our time may be seen in great abundance, coasting along the shores of the North Sea from the Mediterranean to the icy regions of high latitudes ; but hitherto it has not been observed on our side of the equator.

The females are said to carry their young six months, and that the cub at birth measures about 20 inches long.   Mr. Knox, indeed, gives us the particulars of one taken from a mother, killed in the Firth of Forth, of less than 5 feet in length, which actually measured 27 inches, or considerably more than one-third of the entire length of the parent. The young are carefully attended to by their dams ; and it requires about ten years before they attain to full maturity.

In their every-day habits, porpoises resemble greatly the dolphins, being equally fleet and voracious.   Vast troops of them herd together, keeping in shore, in pursuit of the periodic shoals of herrings, mackerel, and other fish, to, as it may be easily imagined, their great destruction.

These animals have been known to take a bait, and some have thus been captured by the hook, although, in most instances, they prove too strong for the line.   This kind of fishing reminds one of the giant angling, possibly derived from this source—" He sat upon a rock, and bobbed for whale."

---

[1] *communis,* common.
[2] From the French " *porc-poisson,*" hog-fish.   The porpoise is known to British sailors by the names of sea-pig and herring-hog; to the French, *marsuin* ; to the Swedes, *marsvin* ; and to the Germans, *meerschwein.*

Family V. DELPHINAPTERIDÆ.[1]

Without dorsal fin; head rounded, shelving; pectoral fin ovate beak of the skull depressed, tapering, rounded at the end; cervical vertebræ more or less anchylosed (?); mandibular symphysis short; bladebone connected to the collar and breast bones by unusually enlarged spinal processes.

Genus DELPHINAPTERUS,[1] Lacepede.

Teeth small, slender, conical acute; pectoral fin rather slender ; beak of the skull elongate ; palate flat, not laterally grooved behind; bladebone very broad, nearly semicircular ; lower jaw longest.

DELPHINAPTERUS PERONII, Lacepede. Péron's Dolphin.

Synonyms—*Delphinus Péronii*, Lacep.; Cuvier, &c.
Delphinapterus leucorhamphus, Péron.
Delphinapterus Péronii, Lesson ; Jardine, Nat. Lib. ; Gray,
S. & W., p. 276, Suppl. p. 72.

Teeth $\frac{38-38}{38-38}$, to $\frac{40-40}{40-40}$, small, curved inwards.

Colour, upper portion of the body deep black ; the beak, the inferior portion of the sides, including the pectoral fins, with the exception of a brown-black spot on their hinder margins, and the belly, of a pure and dead white ; the two colours distinctly parted from each other. Length, 6 feet to 6 feet 4 inches.

Inhab: South Atlantic ; New Guinea.

Genus NEOMERIS, Gray.

Teeth small, compressed, slightly notched ; pectoral fin ovate, falcate; beak of the skull short ; blade-bone large, triangular.

NEOMERIS[2] PHOCÆNOIDES, Dussumier. The Neomeris.

Synonyms—*Delphinus phocænoides*, Dussum., Cuvier.
Delphinapterus melas, Temminek.
Neomeris phocænoides, Gray, S. & W., p. 306 ; Suppl., 82.

Teeth $\frac{16-16}{16-16}$ to $\frac{20-20}{20-20}$, compressed, oblique.

Colour, black. Length, 4 feet.

Inhab: Indian Ocean, Bay of Bengal, Japan, Cape of Good Hope.

---

[1] *delphinus*, dolphin, and ἄπτερος, without (dorsal) fin.

NEOMERIS (?) BOREALIS. Peale.

Synonyms—*Delphinapterus borealis*, Peale, Expl. Exped., 1848.

Delphinapterus (?) borealis. Gray, S. & W., p. 277.

Black, with a white lanceolate spot on the breast, which is extended in a narrow line to the tail. Length, 4 feet.

Inhab. North Pacific Ocean.

. "While in the water it appears to be entirely black, the white line being invisible. It is remarkably quick and lively in its motions, frequently leaping entirely out of the water, and, from its not having a dorsal fin, is sometimes mistaken for a seal."—*Peale.*

Mr. Cassin remarks of this animal " it appears to us probable that it does not belong to the genus Delphinapterus, or to the group of which D. Péronii is the type." To which Dr. Gray adds—"this species appears to resemble Delphinapterus only in the absence of the dorsal fin, in which respect it also resembles the Beluga, of which it is probably a species."

From the diminutive size and general black colour of this dolphin, I am inclined to place it as a species of the genus Neomeris, that is, if it be not identical with the previous one, their ascertained habitats being within a few degrees of the same parallels of latitude.

I am strengthened in this supposition by the result of Dr. Gray's comparison between two skulls of the N. phocænoides, one from Japan, now in the Leyden Museum, and the other from the Cape of Good Hope, in the Paris collection—"the skulls are very much alike, but they may be two species, characterized by the number of the teeth."

Taking it for granted that the various habitats assigned to these unmistakeable animals have been correctly ascertained, I am impelled to ask, why limit to within a very narrow compass the rovings of the Orca capensis or pacifica, the largest and fiercest of the "gladiators" of the ocean, when such extensive powers of locomotion are yielded without remonstrance to the species of this diminutive group? " I doubt," writes Dr. Gray, " its (the skull of the O. pacifica, obtained by Captain Delville, R.N., from the North Pacific), being from the *North* Pacific, as I believe there is a skull of the same species in the Paris Museum, collected by M. Eydoux, and said to have come from Chili, *South Pacific.*" Why not? This is surely straining at a gnat, and swallowing a camel.

The Delphinapterus Péronii agrees in cranial and dental structure with the common dolphin, while the Neomeris phocænoides presents on these points the characteristics of the common porpoise, and, both in their want of the dorsal fin and in the possession of the large spinal processes of the blade-bone, approach the White Whale of the Northern Seas: the group thus affording a convenient connecting link between the rapacious and the more timid of the cetacea.

(*b.*) Teuthophaga.[1] Squid-eaters.

B. Teeth deciduous, numerous in front part only of both jaws.

Family VI. BELUGIDÆ.[2]

Without dorsal fin; head rounded in front, small, scarcely beaked, pectoral fins small, sub-oval, thick, powerful; skull, very convex, caused by the hinder wing of the cheek-bone bending over the eye-cavity, instead of spreading out horizontally; teeth early, deciduous, conical, oblique, frequently truncated; cervical vertebræ usually free; blade-bone with large spinal processes.

Beluga Catodon,[3] Linnæus.   The Beluga or White Whale.
Synonyms—*Physeter catodon,* Linn.

Beluga catodon, Gray, 1850, S. & W., p. 307, Suppl., p. 94.
Delphinapterus Beluga.   Lacepede.   Scoresby.
The Beluga or White Whale, Jardine, Nat. Lib., vol. 7, p. 204.

Teeth $\frac{8-8}{8-8}$ to $\frac{10-10}{10-10}$, most frequently $\frac{9-9}{8-8}$, conical.

The colour of the adult animal is throughout of a creamy white, whence is derived the name White Whale of the English sailors, and Hvitfisk of the Scandinavian seamen. When young, however, the general colour is much darker, being either of a slaty-grey, mottled with brownish spots, or of an uniform bluish tint; being thus readily distinguishable among the grown-up animals of the herd.

The old attain in length to as much as 20 feet, but the examples usually met with, measure from 13 to 14 feet.

The form of the body somewhat resembles two unequal cones joined at their bases, the shorter one being placed in front. The head is small and lengthened, and the tail thick and powerful. The union of these parts presents a frame replete with graceful symmetry, and well adapted for that velocity for which this species is so celebrated.

The food of the white whale consists of cuttle-fish, large prawns, and of the smaller kinds of fish, such as cod, haddock, &c., so abundantly distributed over the Northern Seas.

Inhab: Higher latitudes of the Northern Seas, principally within the Arctic circle.

The habits of this fine species of whale are decidedly gregarious, commonly frequenting the estuaries of the larger rivers, and very rarely seen far from land. It is found most plentifully in Hudson's Bay, Davis Straits, and in certain portions of the Northern coasts of America and Asia. During the winter large herds exist along the coast of Danish Greenland, but, as summer approaches, they travel towards the north, feeding along the western shores of Davis Straits up to the head of Baffin's Bay, their range of habitat corresponding greatly with that of the Right Whale.

---

[1] τευθος, calamary or squid, and φάγω, I eat.
[2] *Beluga,* from the Russian *Bieluga,* signifying White-fish.
[3] κάτω, under, below, and ὸδούς tooth, in allusion to the teeth being found only in the lower jaw of the sperm whale, which the beluga, but only in some other respects, resembles.

Their blast, according to Mr. Brown, is not unmusical, and, when under the water, they emit a peculiar whistling sound which might be mistaken for the whistle of a bird, and on this account the seamen often call them *sea-canaries*.

The white whale is eagerly sought for by the natives of these inhospitable climes, and many are captured by strong nets spread across the entrance of inlets, or partially so, of the various sounds between the numerous islands.

The natives consider this whale as, next to the seal, the most valuable animal in administering to their wants, for there is scarcely a single part of the body which they do not convert to some beneficial purpose.

The flesh in the fresh state affords savoury and nutritious food for their immediate wants, and when dried, an ample store of supply for the long and dreary winter; the oil, although not abundant, is of the best and finest quality, suitable for drinking, cooking, and burning; the finer portions of the internal membranes are used, instead of glass, for glazing the windows, and the coarser for bed furniture; the skin, duly prepared, offers an excellent substitute for leather; and the sinews furnish strong thread and string for the purposes of sewing or fastening their various utensils.

The whalers rarely kill the Beluga, because the swiftness, agility, and resistance displayed by it, when struck, cause more trouble than the yield of oil is worth.

Although the Beluga migrates towards warmer waters during the intense severity of the Arctic winter, only a few stray wanderers have, as yet, been seen so far to the south as the Frith of Forth.

Dr. Gray considers the following, only as varieties of the B. catodon :—

Beluga rhinodon, Cope. Arctic Seas.
Beluga declivis, Cope. Artic Seas.
Beluga angustata, Cope. Arctic Seas.
Beluga Canadensis, Wyman. Canada.

BELUGA KINGII. The Australian Beluga.

Synonyms—*Delphinus (Delphinapterus)*, Gray, 1827.

       Beluga Kingii, Gray, S. & W., p. 309, Suppl. ; p. 95.

Teeth $\frac{10\text{-}10}{9\cdot9}$, small, hooked.

Inhab. Coasts of New Holland.

The skull of this dolphin was presented, in 1826, forty-six years ago, to the British Museum, by the late Admiral King, and up to the present period no further discoveries of its existence have been made. It is, therefore, quite sufficient to record that a peculiar whale, supposed to belong to the genus Beluga, still lives in our seas, and that the acquisition of other examples would be desirable to confirm, or otherwise, its generic character.

Family VII. GLOBIOCEPHALIDÆ.[1]

Head much swollen, globe-shaped; forehead very prominent; beak scarcely visible; dorsal fin falcate, central; pectoral fins long, narrow, placed near each other on the chest; skull broad across supra-orbital ridge; intermaxillary bones very wide; beak slightly over half of the entire length of skull; mandibular symphysis, very short (⅓ length of ramus, G. macrorhynchus); cervical vertebræ, anchylosed; blade-bone triangular, with large spinal processes; teeth conical, large, early deciduous.

Genus GLOBIOCEPHALUS,[2] Lesson.

" Skull,—palate flat; beak rather tapering in front."—Gray.

GLOBIOCEPHALUS[1] MELAS, Traill. The Deductor,[2] or Caa'ing[3] Whale.

Synonyms—*Globiocephalus melas*, Traill.

> *Cachalot svineval*, Lacepede, 1804.
>
> *Delphinus globiceps*, Cuvier, 1812.
>
> *Delphinus deductor*, Scoresby, 1820.
>
> *Globiocephalus deductor*, Lesson, 1827. Jardine, 1843.
>
> *Grampus globiceps*, Gray, 1828.
>
> *Globiocephalus svineval*, Gray, S.&W. p. 314. Suppl. p. 83.
>
> *The Caa'ing Whale*, Neill, 1836.

Teeth, $\frac{11\text{-}11}{11\text{-}11}$ to $\frac{12\text{-}12}{12\text{-}12}$, rarely $\frac{14\text{-}14}{14\text{-}14}$, slightly curved at their tips.

Inhab.: North Sea.

Colour, smooth and shining jet-black on the upper parts of the body, somewhat paler underneath, relieved by a white streak from the throat along the abdomen. The length varies from 6 to 26 feet; the pectoral fins are very long, 6 to 8 feet each, and narrow, being the reverse to the exceedingly broad swimming paw of the O. gladiator.

The food of the deductor is similar to that of the preceding animal, namely, cuttle-fish, crustaceans and small fish; and when the aliment is plentiful, it becomes exceedingly fat, and affords a large quantity of excellent and valuable oil.

In their habits the members of this family appear to be the most sociable of the cetacea, herding together in large flocks; and in their disposition very timid and wholly inoffensive. " In all instances on record of their being discovered at sea, and hunted to land, the chase has been free from danger, and a few frail boats and most ineffective weapons, with shouts and noise in the water, were sufficient to drive them from their native element to their destruction."

---

[1] *globus*, a globe, and κεφαλή, head.
[2] *deductor*, a follower (when one is driven ashore the rest follow).
[3] From the scotch, *caa*, signifying to drive, being the ordinary method of their capture, viz., by driving them ashore.

These gentle animals, when excited into a state of confusion and alarm, are highly characterized by an overpowering instinct which compels them to follow any leader without reference to age or experience, although by so doing they are led into certain danger. The nature of this impulsive force may be familiarly exemplified by the similar habit entertained by sheep : thus, when a flock is required to enter into any strange place of confinement, or to pass along the narrow stage in order to reach the deck of a vessel, it is necessary that one of them be caught and slowly dragged to its destination, the louder the bleating the better, when the whole lot will precipitately follow; so, with this species, if one individual be wounded and take the ground, the others will blindly rush on to their fate; or, rather, as in the method of stranding a shoal, "the men engaged in the hunt, at first succeeded in stranding only one of the young cubs; it immediately set up loud cries, which were heard by the rest, and an old one, probably its mother, speedily came to its relief; but she came not alone, the whole flock followed, and were thus made an easy prey."

GLOBIOCEPHALUS MACRORHYNCHUS,[1] Gray.   The South Sea Black-fish.

Synonyms—*Globiocephalus macrorhynchus*, Gray, S.&W. p. 320; Suppl. p. 84.

Black-fish of the South Sea Whalers, Bennett, Wh. Voy. p. 233 ; Crowther, P.Z.S., 1864.

Teeth $\frac{8-8}{8-8}$, subcylindrical, variable in number.

Colour, black.   Length, from 16 to 20 feet.—" Head, thick, square, and short; the snout blunt, and but little prominent.   The angles of the lips are curved upwards, giving the physiognomy an innocent, smiling expression."—Bennett, Whaling Voyage, p. 233.

The museum skulls of this species correspond closely in every characteristic formation and in size with the particulars given of those of the G. melas, and I feel incapable to point out any salient character by which to separate the species.

These animals are abundant in our seas, and, according to Mr. Bennett, roam about the ocean in very large troops, and appear to inhabit the greater portion of the aqueous globe, uninfluenced by the remoteness or vicinity of the land.   He observed examples in many parallels of latitude between the equator and 50° N. and 35° S., in the central part of the Atlantic and Pacific Oceans, as well as off the coast of California, and in the Indian Archipelago.   " Sperm whalers," continues Mr. Bennett, " often attack this species with their boats, in order to obtain a supply of oil for ship consumption; some risk, however, attends their capture, for when harpooned, they will sometimes leap into a boat.   A Blackfish of average size will produce from 30 to 35 gallons of oil, which in its most recent state has a dark colour and an unpleasant odour."

[1] μακρός, broad, and ῥύγχος, beak.

Dr. Crowther more recently communicates, in the Proc. Zool. Soc. London, the following information, which materially adds to our knowledge of the habits of this cetacean :—"This species is in reality a miniature Sperm Whale in its habits, &c., feeding upon the same food, geographically occupying the same localities as the Sperm Whale, following the great equatorial currents, so long as they retain their warmth, and met with in the greatest numbers in the southern hemisphere at those points where the equatorial meet the polar currents, eddies being formed in which no doubt the squid collects. I am not aware that Blackfish preys upon anything but squid[1] ; it is essentially gregarious, countless hordes being met with where food is abundant. Length, 12 to 15 feet; diamater, 3 to 4 feet; weight, 2 to 3 tons, the former about the average. Oil, the only kind that will mix with sperm."

Several other animals, said to be of the genus Globiocephalus, have been claimed as distinct species by zoologists ; they may be so, but no distinctive quality of any importance has been pointed out, beyond those variations incident to every one of the cetacea, and perpetually exhibited among the best known and most familiar of the family.

These are—

G. affinis, Gray. North Sea.
G. intermedius, Harlan. Coasts of North America.
G. Edwardsii, A. Smith. Coasts of Cape of Good Hope.
G. Grayi, Burmeister. Buenos Ayres.
G. Scammonii, Cope. North Pacific.
G. Indicus, Blyth. Bay of Bengal.
G. Sieboldii, Gray. Japan.
G. Sinensis, Blyth. China.
G. Sibo, Gray. Japan.

The foregoing list is compiled from Dr. Gray's Supplement to the Seals and Whales. pp. 84, 85.

Genus SPHÆROCEPHALUS,[1] Gray.

" Palate of the skull convex ; beak oblong, of nearly the same width the greater part of its length." Gray.

SPHÆROCEPHALUS[2] INCRASSATUS,[3] Gray. The Thick-palated Deductor.

Synonym—*Sphærocephalus incrassatus*, Gray, S. & W., p. 324 ; Suppl. p. 85.

Teeth $\frac{9\text{-}9}{0\text{-}0}$ or $\frac{10\text{-}10}{0\text{-}0}$.

Inhab : British Seas.

---

[1] Dr. Crowther writes to me, that subsequently to this assertion, ho finds that in addition to the squid, the black-fish consume large quantities of fish and crustacea ; for in the agonies of death, conger eels, crabs, crayfish, &c., have been ejected from their stomachs.

[2] σφαῖρα, a globe, and κεφαλή, the head.

[3] Thickened.

A single skull in the British Museum has enabled Dr. Gray to create the foregoing genus :—" It is so distinct, both in the form of the nose of the skull, in the width of the maxillary bones, and more especially in the thickness and convexity of the palate of the front part of the skull, from the species which have hitherto been described, and the differences are so visible, that Mr. Edward Gerrard selected it as a distinct species, as soon as he saw it."

C. Long produced spiral tusk in upper jaw of male only.

Family VIII. MONODONTIDÆ.[1]

Without dorsal fin ; head short, rounded in front, scarcely beaked ; pectoral fins sub-oval, much longer than broad ; skull very convex, the hinder wing of the cheek-bone bending over the eye-cavity ; blade-bone with large spinal processes ; cervical vertebræ, usually free ; costo-sternal ribs ossified ; teeth few in both jaws, early deciduous, with the exception of one in the left side of the upper jaw of the male, occasionally in the female, developed into a very long spiral tusk, projecting forwards in a line with the axis of the body.

Genus MONODON,[1] Artedi.

The characters descriptive of the genus, being the only one of the family, are given above.

Mr. Flower remarks that the Monodon and Beluga are, in almost every part of their skeleton, nearly identical, and he considers the exceptional dentition of the former as an aberration of secondary importance ; he therefore unites the two genera into a distinct sub-family, placing it next to the Platanistidæ.

MONODON MONOCEROS,[2] Linnæus. The Narwhal, or Sea Unicorn.

Synonym—*Monodon monoceros*, Linn., Schreb., Desm., Scoresby, &c., Gray, S. & W. p. 310 ; Suppl. p. 95.

> *Sea Unicorn*, Sowerby.
> *Narwhal*, Blumenb., Klein.

The colour of this singular animal is dusky black on the upper surface, greyish on the sides and white underneath, variegated at different stages of its existence with more or less darker streaks and patches, disposed more numerously on the sides. The food of the Narwhal consists of cuttle-fish, crustaceans, fish, &c., and Mr. Scoresby records the contents of the stomach of one, killed by his crew ; " they consisted of several half-digested fishes, with others of which only the bones

---

[1] μόνος, one, and ὀδούς, tooth.
[2] μόνος, one, and κέρας, a horn, a tusk.

remained. These were the remains of a cuttle-fish, part of the spine of a flat fish, probably a small turbot, and a skate almost entire. The last was 2 feet 3 inches in length, and 1 foot 8 inches in breadth. It comprised the bones of the head, back and tail, the side fins, and considerable portions of the muscular substance. It appears remarkable that the Narwhal, an animal without teeth, a small mouth, and with stiff lips, should be able to catch and swallow so large a fish as a skate, the breadth of which is nearly three times as great as the width of its own mouth."

The distinctive character of the Narwhal, not being possessed by any other whale, lies in the long projecting spiral tusk, produced probably by the excessive growth of the canine tooth. This formidable weapon has been known to attain as much as 8 or 10 feet in length, while that of the animal, of course exclusive of this appendage, was from 14 to 16 feet only. It is hollow at the base and solid towards the extremity, and composed of fine close-grained ivory, of a dazzling and enduring whiteness, so extremely hard as to take a high polish. Formerly these tusks, very rarely brought to Europe, were regarded as the veritable horns of the fabulous unicorn, and were consequently valued as inestimable and almost priceless curiosities. The establishment of the Greenland fishery quickly dispelled all doubts as to the nature of their real character, and the present value now depends solely on the number of pounds weight the tusk might weigh.

The Narwhals are gregarious, and met with in considerable numbers in the numerous creeks and bays of Greenland, Davis Straits, and Iceland, but solitary individuals occasionally stray as far south as the northern parts of Great Britain.

Tusks of the Narwhal, in a semi-fossilized state, have been found in Siberia, on the coast of Essex and of Lyons.

D. Teeth in lower jaw only.

Family IX. GRAMPIDÆ.[1]

Head rounded, somewhat obtuse; forehead very convex, not so protuberant as in Globiocephalus, scarcely beaked; dorsal fin distinct, but low; pectoral fins well developed, ovate, rather elongate, placed low down on the side of the body; skull depressed; symphysis of lower jaw short; cervical vertebræ anchylosed; costo-sternal ribs ossified; sternum composed of one piece, broad in front; teeth few, conical in front part only of lower jaw; those of the upper jaw early deciduous.

The term *Grampus* (*great fish*) is, and has been, in scientific works, and in general conversation, very universally applied to denote among the odontocete the formidably dentated animal, the Killer; and that of *Blackfish* to distinguish the cetaceans of milder propensities and of

---

[1] *Grampus*, contracted from the French, *grampoise, grand poisson, great fish.*

greater usefulness to man, such as the sperm, the caa'ing, and some other whales. Dr. Gray's present arrangement of the grampidæ abruptly ignores this common understanding among people of many nations, and brings together under the old familiar name a group of beings, whose every trait of character is of exactly a contrary nature to that of the savage gladiator, and to whom the word Blackfish would have been much more suitable. In looking over the Catalogue of Seals and Whales, I find therein the following generic names: *Hunterius, Macleayus, Eschrichtius, Cuvierius,* and *Sibbaldius;* why not, in order to restore the Grampus to its original standing, and also as being appropriate, substitute the generic name now employed for that of *Grayius,* in honour of one distinguished in every branch of zoology, but more particularly so in this, the marine mammalia?

GRAMPUS (GRAYIUS?) GRISEUS, Cuvier. Cuvier's Dolphin.

Synonyms—*Phocœna grisea.*—Lesson.
    Delphinus griseus.—Cuvier.
    Grampus Cuvieri.—Gray. S. & W., p. 295; Suppl. p. 83.
Teeth $\frac{0-0}{2\cdot2}$, truncated.

Colour, bluish-black along the upper portions of the body, gradually assuming a dull white on the abdomen. The length seldom exceeds ten feet.

The habit of herding together, of following a leader, and of uttering cries when stranded, appears to be possessed by this cetacean, in common with the globiocephalus; but so much confusion by the indiscriminate use of the vague term "blackfish" is caused, that it is almost impossible to define with any degree of certainty what species of animal is meant.

Inhab : North Sea—Coast of Hampshire, England.

GRAMPUS (GRAYIUS) RISSOANUS, Laurillard. Risso's Dolphin.

Synonyms—*Delphinus aries,* Risso.
    Delphinus Rissoanus, Laur.
    Delphinus de Risso, Cuvier.
    Grampus Rissoanus, Gray, S. & W., p. 298, Suppl. p. 82.
Teeth $\frac{0-0}{3\cdot3}$ to $\frac{0-0}{7\cdot7}$, rather small, truncate.

Colour of the body bluish-white, relieved with irregular brown-edged scratch-like lines in all directions. Females uniform brown, with similar scratches.

This species is very nearly allied to the preceding one, being of the same form and size, but differs principally in a slight variation of the colouring of the body, and the habitat of the animal—considerations of minor importance in distinguishing species, and frequently deceptive.

Inhab: Nice, Mediterranean.

GRAMPUS (GRAYIUS) (?), RICHARDSONII, GRAY. Richardson's Dolphin.
Synonym—*Grampus Richardsonii*, Gray, S. & W., p. 299, Suppl. p. 83.

Teeth, $\frac{0\,0}{4\cdot4}$, rather large, far apart, sub-cylindrical at the base, symphysis of lower jaw wide in front.

Dr. Gray describes this species from the lower jaw only, which presents in the symphysis a variation of structure sufficient in his opinion to constitute a new species.

Inhab: Cape of Good Hope.

Family X.   HYPEROODONTIDÆ.

Head beaked; cheek-bones and hinder edge of the skull greatly elevated into high occipital crests by vertical bony partitions, rendering the form of the skull different to that of any other cetacean, and highly asymmetrical; upper jaw larger and broader than the lower one, dorsal fin small, subfalcate, placed much beyond the middle; pectoral fins small, ovate, low down on the sides of the body; costo-sternal ribs cartilaginous; mandibular symphysis rather long.

Teeth, $\frac{0\cdot0}{1\cdot1}$, or $\frac{0\cdot0}{2\cdot2}$, well in front, small, conical, frequently hidden in the gums.

Genus HYPEROODON,[1] Lacepede.

Forehead rounded in front; beak of the skull with a high crest on each side, and with maxillary tuberosities at its base; cervical vertebræ firmly anchylosed.

HYPEROODON[1] BUTZKOPF[2], Bonnaterre.   The Bottle-head.

Synonyms—*Delphinus butzkopf*, Bonn.

Hyperoodon butzkopf, Lacep., Gray, S. & W., p. 330, Suppl. p. 97.

Hyperoodon rostratus, Lilljiborg, Flower.

Inhab : North Sea.

The colour of this frequently-caught whale on the English coast is glossy-black, becoming a pale lead-colour underneath. It is a large animal, reaching to the length of from twenty to thirty feet. The food consists principally of various kinds of cephalopods, for in the stomach of one stranded in 1853, were many hundred cuttle-fish beaks, so placed one within the other as to ride regularly imbricated in rows of ten, fifteen, or twenty together.

---

[1] ὑπερ, above, and ὀδούς, tooth ; in allusion to the teeth in the palate described by Baussard, but which in all other specimens have never been seen.
[2] *butzkopf* (German), pointed head.

HYPEROODON LATIFRONS[1], Gray.   The Heavy-headed Hyperoodon.

Synonyms—*Hyperoodon latifrons*, Gray, Voy. Erebus and Terror.
   Lagenocetus latifrons, Gray, S. & W., p. 339. Suppl. p. 97.

Inhab : North Sea.

The reflexed parts of the cheek-bones are in this species much thickened above, and in their altitude exceed the hinder edge of the skull; lower jaw straight, and also the beak of the skull.

Reinhardt remarks that "Eschricht believed, as is known, that Hyp. latifrons was established on a very old male of the common Dögling (Hyperoodon rostratus), but Gray's species must now be regarded as well grounded."

Genus ZIPHIUS, Cuvier.

Forehead tapering; beak of the skull simple, and without tuberosities at its base ; respiratory aperture deep-seated ; cervical vertebræ partially anchylosed.   The intermaxillaries at their base, and the occipital bones, form by their enlarged prominent edges around the sides and behind the brain-case a large hemispherical cavity, which serves to receive the head matter or spermaceti.   The cranium thus exhibits a strong connecting link in its general features between the Hyperoodon and the Physeteridæ; with the former, by the elongated beak and almost edentulous condition of the jaws, and with the latter by the well defined spermaceti-cavity.[2]

ZIPHIUS CAVIROSTRIS[3], Cuvier.   The Mediterranean Ziphius.

Synonym—*Petrorhynchus mediterraneus*, Gray, Suppl. p. 98.

Inhab : Mediterranean.

The skull of this singular cetacean was described about half a century ago by Cuvier as the type of an extinct species, from the semi-fossilized state in which it was found ; but which view, M. Gervais, from other examples has shown to be erroneous.

ZIPHIUS GERVAISII, Duvernoy.   Gervais' Ziphius.

Synonyms—*Hyperooden Gervaisii*, Duvernoy.
   Ziphius Gervaisii, Fischer.
   Epiodon Desmarestii, Gray, S. & W., p. 341, Suppl., p. 98.

Inhab : Mediterranean.

The only notice we possess of this cetacean is derived from the short account by Risso, of the form and colouration of the living animal, presuming his species to be the same as that described by Fischer, but of which I am uncertain, nay, very doubtful.

---

[1] *latus*, broad, ample, and *frons*, forehead.

[2] Of the *Ziphius cavirostris*, Cuvier observes—" *Cette tête a, comme on voit, de grand rapports avec le cachalot, et encore de plus grands avec l'Hyperoodon,*" Vol. v, p. 351, 1825.

[3] *cavus*, hollow, and *rostrum*, the beak.

"Steel-gray, with numerous irregular white streaks; beneath white; body thicker in the middle; tail slender, long, keeled; rounded on the belly; head not swollen, ending in a long nose; upper jaw short, toothless; lower jaw much longer, bent up, and with two large conical teeth at the end; teeth nicked near the tip; the eyes small, oval; blowers large, semilunar; pectoral fins short; dorsal fin rather beyond the middle of the back; the caudal fin broad, festooned. Length, nearly 16 feet. Inhab: Nice; common, March and September," from Dr. Gray.

ZIPHIUS INDICUS, Van Beneden. The Cape Ziphius.

Synonyms—*Ziphius Indicus,* Van Beneden.
Ziphius du Cap-de-Bonne Espérance, Gervais.
Petrorhynchus Capensis, Gray, S. & W., p. 346, Suppl. p. 98.
Inhab: Coasts of Cape of Good Hope.
A well-defined species, ascertained from skulls in the Paris and British Museums, and likewise from fossil remains found in the red crag beds of England of the Pleiocene period, the strata of which contain a great number of examples of extinct species of this and other families, and a few only of recent forms.

Genus CHONEZIPHIUS,[1] Duvernoy.

The intermaxillaries at their anterior extremity are even and, connecting at their tips, display prominently a large channel: they, however, become towards their base very asymmetrical, forming there a deep funnel-shaped cavity.

CHONEZIPHIUS PLANIROSTRIS, Cuvier.

Synonym—*Ziphius planirostris,* Cuvier.
Found in the crag formation, France.

Family XI.   PHYSETERIDÆ.[2]

Head of moderate or excessive size, more or less truncated in front; mouth placed well beneath, rendering it necessary for the animal to turn on its side or back in order to seize its prey, an action not known in any other cetacean; upper jaw toothless[3], broader, more massive, and slightly shorter than the lower one; under jaw slender, cylindrical in front, and received within the pendent upper lips, as in a furrow; armed with numerous teeth, having pulp-cavities at their base; symphysis of

---

[1] χώνη, a tunnel or funnel, and *Ziphius.*
[2] φυσητήρ, blower.
[3] That is, without functional teeth.

lower jaw moderate or excessive in length : pectoral fins short, broad, comparatively small, and weak ; dorsal fin or hump distinct but small ; bones of the skull so raised at their edges as to form on the summit of the head a large basin for the reception of spermaceti ; cervical vertebræ anchylosed into one piece, with the exception of the atlas in the sperm whale ; four anterior pairs of ribs attached to the sternum by unossified cartilage. Males larger than females.

Genus KOGIA, Gray.

Head moderate, short, broad ; forehead elevated ; mouth small, and placed beneath the projecting snout "like that of a shark"[1] ; pectoral fins weak ; dorsal fin small, depressed, rising and falling with the line of the back, at an obtuse angle ; skull very broad, rounded behind ; beak short, flat above, rapidly tapering to a point, and nearly equilateral with the breadth at the supra-orbital ridge ; back part of the spermaceti cavity longitudinally divided into two unequal parts by a sinuous ridge of bone ; lower jaw wide at the condyles, contracting suddenly a little beyond half-way, where it becomes narrow and rounded at the tip ; mandibular symphysis about one-fourth of the entire length of ramus ; cervical vertebræ anchylosed into one piece.

KOGIA BREVICEPS, De Blainville. The Short-headed Whale.

Synonyms—*Physeter breviceps*, De Blainville.

Kogia breviceps, Gray, S. & W., p. 217, Suppl. p. 60.

Teeth $\frac{0 \cdot 0}{13 \cdot 13}$, long, slender, acute, conical, arched inwardly.

This species was founded by De Blainville upon a single skull in the Paris Museum, and which closely resembles in every important particular the skulls of the perfect specimens found in the Australian Waters.

Inhabits Cape of Good Hope.

KOGIA GRAYI, Macleay. Gray's Kogia.

Synonyms—*Euphysetes Grayi*, Macleay (Wall), 1851. Gray, Suppl., p. 392.

Physeter simus, Owen, Trans. Zool. Soc.

Kogia Grayi—Gray, S. and W., p. 218.

Kogia Macleayi—Gray, Suppl., p. 391.

Euphysetes Macleayi—Krefft, Proc. Zool. Soc., 1865.

Teeth $\frac{0 \cdot 0}{13 \cdot 13}$, long, slender, arched inwardly.

From pages 39 to 42 of the "History and description of the skeleton of a new Sperm Whale, lately set up in the Australian Museum by Willm. S. Wall, Curator, together with some account of a new genus of Sperm Whales, called Euphysetes, Sydney, 1851,"—of which publication the author was the late eminent zoologist Mr. William S. Macleay,—I select a few paragraphs admirably descriptive of the skeleton of this species.

---

[1] Krefft, Proc. Zool. Soc., Nov., 1865.

"Like a dolphin, it had a low snout, and rising from it a convex forehead, at the base of which was the large single blow-hole, placed at about the middle of the head. The snout was turned up with a margin somewhat like that of a pig. In the gums of the roof of the mouth there was on each side a series of sockets for receiving the teeth of the under jaw; these teeth were hollow, conical, and inserted somewhat horizontally in the sides of a very thin, narrow, sub-cylindrical under-jaw. The eye was situated low, in front of a very weak pectoral fin. There was a triangular dorsal fin like that of a dolphin, the rather convex front edge of it being inclined backwards at an angle of 45°. The hinder edge of it was more perpendicular and concave. The perpendicular height of the point of this dorsal fin from the back was about 3½ inches, and its base 6 inches wide.

"There is the same want of symmetry, the same distortion of the bones, and the same concavity of the upper surface of the head, formed by the enormous development of the base of the maxillaries, and the same convexity of the roof of the mouth, as are found in the genus Catodon.

"The lower jaw is a singular contrast to the upper, the former being as slight and fragile as the latter is massive and strong. So weak is the connection of this under-jaw with the skull, that the articulating condyles are scarcely to be detected. The broad branches are nearly as thin as paper, and although the sides are reflexed inwardly, as in dolphins, the doubling, so as to form the hollow tube, does not occur as in them, near the base of the jaw, but within three inches of the symphysis. Nevertheless, so extremely feeble an under-jaw demonstrates that the long, sharp teeth serve merely for the purpose of retaining the weak mollusca which, no doubt, forms this creature's prey."

This whale, in length between 9 and 10 feet, was stranded, in 1850, on the Maroobrah Beach, half-way between Coogee and Botany, and the imperfect remains, collected by Mr. Wall under great difficulties, are now set up in our Museum, presenting an interesting, although indifferent, specimen of a species previously unknown, except through the medium of the single skull in France.

Fifteen years subsequent to this discovery, another specimen, but of larger dimensions and happily in perfect condition, was cast ashore on the beach at Manly, which afforded Mr. Krefft an excellent opportunity for examining the external form, taking the various admeasurements of lengths and girths, and for securing an almost complete series of the bones, necessary for the erection of the artificial skeleton—operations greatly enhanced in value by photographic illustrations of the external form of the living animal, and a few of the essential adjuncts.

The Manly Beach animal and that from Coogee correspond so greatly with each other that I have no doubt as to their specific identity; nevertheless they exhibit certain differentiating characters worthy of remark.

ton of the last found example the rami of the lower
a much thicker substance; the teeth with which
) longer, stronger, and partly incurved; the edges
: are more rounded off; and the vertebræ greater
hibited in the relative parts of the whale described
he first three deviations may reasonably be attri-
effects of age, and the last one to the loss, together
nall bones, of the lesser caudal vertebræ of the

fin, placed far behind the middle of the body, in
:n can claim no more appropriate appellation than
mbling that of the sperm whale, whereas in the
:sal is described like that of a dolphin; but, it must
: this description was taken from a much torn and

1otice of this whale, published in the Zool. Proc. of
tal length at 10 feet 8 inches, and the colour as
owish beneath, and he considers it a distinct species
found.
lian coast.

GENUS PHYSETER,' Linnæus.
:onical, set wide apart, only in lower jaw; head
equalling nearly one-third of the entire length of
runcated in front; nostrils disproportionate in size,
,e largest; nasal and facial bones generally unsym-
:d; no true dorsal fin, but in lieu a distinct dorsal
and the tail an irregular, ridge-formed protuberance,
hree very small fins; gullet capacious, in direct
he gigantic Right Whale; cervical vertebræ firmly
: exception of the atlas, which is free; lower jaw
tive when compared with the bulk of the head;
excessively long, nearly two-thirds of the entire
; ribs comparatively slender, their tissue, like that
.se and compact; anterior four pair of ribs attached
unossified cartilage; sternum composed of three
e crown of the head immense, protected externally
;ument, and divided internally into cells by a similar
ntain that peculiar oil, liquid when recent, but soon
:te or granulated form known as spermaceti.
ch larger than the females.

---

remains of a cetacean, apparently dead for about six weeks)
1ad been so much devoured by native dogs and other animals
nained of the external integuments except the flukes of the
,humb extremity of the right pectoral fin, the fore part of the
: gums, and part of the under-jaw with the teeth and lip
are all much torn, &c." p. 37.
)w the wind, a pair of bellows—in allusion to the structure of
:apable of throwing up jets of spray.

PHYSETER MACROCEPHALUS,[1] Linnæus.   The Sperm or Spermaceti Whale.

Synonyms—*Physeter macrocephalus*, Linn., O. Fabr., Shaw, Bell, Flower, Murie.

Catodon macrocephalus, Lacep., Gray, S. & W., p. 202, Suppl., p. 59.

Catodon australis, Gray, S. & W., p. 206.

Meganeuron Krefftii, Gray, S. & W., p. 387, Suppl. 59.

Teeth $\frac{20}{25}$ to $\frac{20}{27}$ on each ramus, large, conical, often worn down.

The above generic characters apply to this species, and which I believe to be the only one of the true sperm whale.

The massive head of the living animal is truncated in front; the blunt extremity, which projects considerably beyond the lower jaw, is composed of a thick, fatty substance, called by seamen the *junk :* this junk is at all times large, but in a large whale it weighs between two and three tons, and serves from its elasticity to act as a buffer to lessen the effect of sudden concussion accidentally received when the huge mass is in rapid motion.

The great weight of the bones of the skull is efficiently buoyed up, partly by the light material of which the junk is composed, but principally by the large quantity of oil, of course of much lesser specific gravity than the surrounding water, pent up in the cranial reservoir, and usually estimated at from three to five hundred gallons ; a fact well illustrated by the few lean individuals occasionally met with at sea, who express by their actions a general want of buoyancy, but more especially in the head.

The mouth is large, lined throughout with a pearly white membrane, and terminates in a throat sufficiently large to allow a free passage to the body of a man.   The eyes are small, and placed over the angle of the mouth.

Under the outer skin lies a layer of yellowish-coloured blubber, termed by whalers the blanket, which varies in thickness from 8 to 14 inches, the breast, dorsal hump, and upper margin of the tail affording the thickest, and from which is obtained the sperm oil of commerce. The more valuable product, the spermaceti, is almost wholly abstracted from the cavity on the crown of the head, and frequently amounts to ten large barrels in the crude state.

The colour of the body along the upper surface is very dark, occasionally black, and fades into a lighter tint on the sides and belly, becoming silvery-grey under the chest.

---

[1] μακρός, long, and κεφαλή, the head.

When fully matured by age, the male of this gigantic race of beings s said to have reached in length to eighty-four feet, with a girth at the largest part of thirty-six feet. In the absence of the actual specimen, an ideal estimate of this enormous bulk may be arrived at by viewing the excellent skeleton, fifty-seven feet in length, of the same species, recently erected in the Australian Museum, and imagining it nearly half again as long.

The ordinary food of the Sperm Whale is derived from various kinds of Calamaries, or squids, and especially from that species called, from its habit of leaping out of water, the flying squid, an animal well known from its extreme abundance in all the open seas of the world, and from its extensive use as bait in the Newfoundland cod fisheries. There can be no doubt, however, that other food, such as fish, crustaceans, and even seals and dolphins, is likewise indulged in.

Otho Fabricius, and other writers of about his time, describe these whales as existing plentifully in the higher latitudes of the northern seas ; but old and young being alike subjected to unceasing persecution, the race has been almost entirely driven away from the North Atlantic. Mr. R. Brown, whom we have had occasion already to quote in regard to the habits of the Northern Killer, remarks of the sperm whale that, " whatever it was formerly, it is now only known to Davis Strait whalers by name, and I could only hear of one recent instance of its being killed on the coast of Greenland near Proven (72° N. lat.), in 1857." Professor Lilljeborg, also, in his " Synopsis of the Scandinavian Whales, 1861," considers the sperm whale as foreign to the Fauna of Norway and Sweden.

There are, however, numerous instances in modern times of their appearance in small groups off the Orkneys, and of individuals being stranded on the British Coast.

" They are essentially inhabitants of the tropical and warmer parts of the temperate seas, and they pass freely from one hemisphere into another. Between the North Atlantic and the Australian Seas there is no barrier interposed to animals of such great powers of locomotion."[1] " Few connect the pursuit of this *sea-beast* with the smiling latitudes of the South Pacific, and the coral islands of the Torrid Zone."[2] Nevertheless they are still the occupants of the colder regions of the south, for " sperm whales were seen in the Antarctic Seas as high as latitude 71° 50' "[3]

Thus, the sperm whale is capable, from its endurance of varied temperature and a permanent supply of food, of roaming at pleasure the entire seas embraced within the Arctic and Antarctic regions.

Were, therefore, the *Fisheries* conducted on judicious principles,—those of capturing the adults at certain seasons only, and at all times sparing the young,—these animals, of the greatest commercial value

[1] W. H. Flower, Trans. Zool. Soc., 1868.
[2] Beale, 1835.
[3] Captain Ross, R.N., Antarctic Voyage.

among the denizens of the deep, would again become as abundant in the waters of both hemispheres as they were in olden times, and afford a steady and profitable source of employment for capital, aud a wide field for the training of a hardy race of seamen.

Although widely dispersed, the cachalot appears to select the strong ocean currents and their back-water as favourite feeding-grounds; for in such places innumerable floating minute molluscs, medusæ and crustaceans are gathered together, and which, in their turn, attract hordes of larger animals, the peculiar prey of our whale.

In habits the cachalots are gregarious, mostly seen in groups, technically termed *schools*, of from twenty to fifty, made up of half-grown males, or of females and their young, guarded by a few of the older males. Large and full-grown males during certain periods go singly in search of food, but when an individual is met with far from the herd, it is usually found to be an *old bull*, who has retired into a solitary state of existence.

When about to emigrate from one feeding-ground to another, several *schools* frequently unite. and conjointly *make the passage*, swimming in a direct course at a rapid rate, with their heads raised well above the water, and their bodies so near the surface that their backs are often seen. Arrived at their destination and replete with food, they become widely scattered, lazily basking on the surface, or in deep repose, or leisurely casting from the nostrils, at each spout, a succession of vapoury jets, at regular intervals of ten or fifteen seconds; or when the fit takes them, gamboling with an uncouth vigorous agility, which frequently displays the entire of the gigantic frame several feet in the air.

The usual rate of speed is from 8 to 10 miles an hour; but the greatest is attained by the painful prick of the harpoon, when it reaches to 15 miles; a velocity very inferior to that of many others of the family, and not to be mentioned as a boast of fast travelling in these railroad times.

Desirous of feeding, or of avoiding an indifferent object, the cachalot *settles down* to the required depth, by gradually and leisurely lowering itself in a horizontal position; when alarmed, the head assumes a downward tendency, and the tail rises vertically in the air, and the animal plunges headlong, almost perpendicularly, into the deep, and remains submerged from three-quarters of an hour to an hour and a quarter.

The male cachalot of 60 feet long, will have a pectoral fin of about 3 feet, and a caudal one, the principal organ of progression, of 19 feet across; in good condition, such an animal will yield about 100 barrels of oil, and 12 barrels of spermaceti.

The female (which rarely attains to one-half of the length of the male), of 35 feet, will measure across the caudal fin 12 feet, produce about 50 barrels of oil, and a proportional supply of spermaceti.

I

The teeth of aged males commonly weigh from 2 to 4 pounds each; the ivory of which they are composed is hard, and capable of taking a high polish; but for commercial purposes it is held in much less estimation than that obtained from the tusks of the elephant.

" The crown jewels of Viti were kept in a wooden box, in charge of the widow of the late Governor of Namose. First, there was a necklace of whales' teeth, the first that ever came to the mountain; secondly, a large whale's tooth, highly polished and carefully wrapped up in cocoanut fibre (whales' teeth are in Fiji what diamonds are with us); thirdly, a cannibal's foot in the shape of a club, and bearing the name of *strike twice*, that is, first the man and then his flesh."[1]

The rare and valuable substance known as ambergris is produced, as a morbid concretion, in the intestines of sickly or diseased cachalots, and usually found floating in impure masses on all the seas of warm climates, or thrown upon the beach; it is then of a greyish colour, mottled with black, somewhat hard and brittle, and when heated, emits a strong, fragrant, musky odour; these lumps occur from a quarter of a pound to forty in weight, and the retail price, of course freed from all impurities, is about one guinea the ounce. Ambergris is now only used as a perfume, its medicinal virtues having long fallen into disrepute.

To those of my readers who take an interest in the details connected with whalers and whale ships, I recommend the perusal of the entertaining descriptions given by Sowerby, Bell, Beale, Bennett, and Jardine, all of whom enter largely, beyond the scientific portions, into many interesting anecdotes relative to the capture of these cetaceans, and to the hardships occasionally endured by the men engaged in their pursuit.

These hardships undoubtedly arise from the protracted servitude on board, inclement seasons, and frequent shipwrecks.

I take no pleasure in recounting the sufferings of these harmless creatures, nor do I especially admire *daring* of men when subjected to no danger, except that caused purely by negligence or accident. I do not consider, as amusement, the many acts of cruelty recorded in these and similar works, unnecessarily committed by sailors, when wholly unrestrained by wholesome enactments.

In conclusion, I cannot refrain from offering a passing tribute of thankfulness that, in my own time, other grand products, profusely derived from inorganic matter, and of an infinitely superior and more economic character, have been discovered to provide for the illumination of our streets and dwellings and for lubricating our machinery, and which have already had a direct tendency to stay the cruel hand, and reduce the waste of life to within much narrower bounds.

I allude to gas and kerosene, which, when supplemented by steel, are steadily superseding in many branches of our industry the employment of whale-oils and whale-bone.

---

[1] Viti: an account of a Government Mission to the Vitian or Fijian Islands, in the years 1860–1861. By Berthold Seeman.

The lessened demand for these still important articles has consequently materially decreased the number of ships engaged in the trade; and, in combination with the diminution of the species, has rendered the returns too precarious for the profitable investment of capital.

I will now briefly notice that portions of the fossil organic remains of the Cachalot have frequently been found, greatly resembling in structure the existing animal.

Professor Owen describes some of these fossil bones, which were obtained from the coast of Essex, England; M. Gervais has named the animal whose relics were discovered at Montpelier, France, in the most modern of the tertiary deposits, the Phys. antiquus; and M. Jaeger mentions, under the name of Phys. molassicus, another species found in Germany.

Genus BALÆNODON,[1] Owen.

Fragmentary relics disinterred from the red crags of the Meiocene period at Felixstowe, England, exhibited teeth very similar to those of the Sperm Whale, upon which character the present genus was founded. M. Meyer has since discovered a skull at Lintry, in Austria, which he places under the Balænodon of Owen, although he thinks that in many particulars, other than the teeth, it approaches nearer to the Zeuglodon than to the Cachalot.

Balænodon physaloides of Owen, and Balænodon lentianus of Meyer, are the two species alluded to.

Family XII. MESOPLODONTIDÆ.

Dorsal fin small, subfalcate; head beaked; forehead receding; throat longitudinally plaited (?); pectoral fins small, low down towards the middle of the chest; skull small, narrow, upper part asymmetrical; frontal portion high; occipital scarcely rounded, flattish; anterior surface of the premaxillæ curves forwards over the breathing apertures; beak much elongated, tapering, narrow; maxillary bones simple, expanding horizontally over the orbits, without tuberosities at the base; inter-maxillaries somewhat swollen behind, not forming a basin round the nostrils; upper jaw shorter and narrower than the lower one, so that when the mouth is closed the upper beak is let within the teeth of the lower one, departing, in this particular, widely from other toothed whales; lower jaw broad behind, narrowed in front; mandibular symphysis moderate, short; cervical vertebræ partially anchylosed; costo-sternal ribs cartilaginous; teeth, at the most, two pair, compressed, in lower jaw only, occasionally largely developed.

---

[1] φάλαινα, balæna, whale, and ὀδούς, tooth.

Genus MESOPLODON, Gervais.

Beak of the skull nearly five-sevenths of the entire length of the cranium, keeled on each side; brain-cavity small; teeth $\frac{0.0}{1.1}$, placed nearly in the centre of each ramus,—of the male large, of the female much smaller; mandibular symphysis about two-sevenths of the entire length of ramus.

MESOPLODON SOWERBIENSIS, de Blainville. Sowerby's Ziphius.

Synonyms—*Physeter bidens*, Sowerby.
    Delphinus Sowerbiensis, de Blainville.
    Heterodon Sowerbyi, Lesson.
    Ziphius Sowerbienis, Gray, S. & W., p. 350, Suppl. p. 101.
    Mesoplodon Sowerbiense, Gervais, Bened.
    Diplodon Sowerbiense, Gervais.
    Mesoplodon Thomsoni (?) Krefft, MSS.

Teeth $\frac{0.0}{1.1}$, much compressed, placed on the anterior third of the ramus, their points directed upwards, and somewhat backwards.

Inhab: North Sea, Coasts of Europe, Coast of New South Wales (?)

Very few solitary specimens of this species in the living state have only been secured since its first discovery in 1800 on the coasts of Scotland, and these have been found stranded on the shores of Ireland, France, Norway, and the Netherlands, to which list of localities may possibly be added that of New South Wales.

Of two of these captured animals, one is described as black above and greyish beneath, and the skin presented a soft, satiny appearance; the other, as having the upper portion of a brownish-lead colour, and the belly bluish and ash.

In length the adults varied from 11 to 16 feet.

The skeleton in the Australian Museum, which, for the present is considered as a synonym, is that of an animal stranded at the latter end of 1870 on the beach near Little Bay, shortly to the north of Botany Heads.

This skeleton I have compared with the excellent engravings of the Mesop. Sowerbiensis in MM. Van Beneden and Gervais "Ost. des Cétacés," but I cannot detect any essential difference of structure between them, although the separating geographic range of habitat is of a maximum nature. I have been lately told that Mr. Flower, on being supplied with a brief description and photographs, has expressed a similar opinion of their identity.

A more careful investigation into details may possibly reveal some differentiating character, and, if I am permitted the use of a little special pleading, " which, together with the great improbability of the same species being found in such widely different regions," may justify its separation from the M. Sowerbiensis. If so, I would suggest that the specific name of Thomsoni, Krefft, be retained, in regard to the memory of one whose loss I consider as a public calamity to this country.

MESOPLODON LAYARDII, Gray. Layard's Ziphius.

Synonyms—*Ziphius Layardii*, Gray, S. & W., p. 353.

Dolichodon Layardii, Gray, Suppl. p. 101.

Inhab. Cape of Good Hope.

Teeth $\frac{0-0}{1-1}$, "in the middle of the sides of the lower jaw. Teeth of the male very long, strap-shaped, produced, arched obliquely, truncated at the end, with a conical process on the front of the terminal edge. Lower jaw weak, very slender in front. Symphysis elongate." Gray.

This singular cetacean is only known from the solitary specimen of a skull in the British Museum ; and the striking peculiarity which it at once presents to observation consists in the elongated teeth of the mandible, for these " arch over the outer surface of the upper jaw, and thus prevent the animal from opening its mouth beyond a very limited extent."

It has been suggested that this strange dental growth, certainly unique, if natural, in the history of living beings, and threatening to prove ultimately fatal to the very existence of its possessor, might have been the result of individual peculiarity, or malformation ; but Dr. Gray thinks otherwise, and has recently formed the genus Dolichodon, or long-toothed, for its reception. It is very desirable that several other examples in a similar state of dental perfection should be brought to light, for otherwise there is nothing to excite surprise.

Genus DIOPLODON,[1] Gervais.

Skull high, narrow, nearly flat behind ; brain-cavity very small ; beak depressed, much elongated, tapering to a point; much narrower and shorter than, and received within the teeth of, the mandible; lower jaw broad behind, contracted in front ; rami high on the sides, rather stout, terminating upwardly in front of the teeth in an arched manner; symphysis short, about one-fifth of the entire length of the ramus ; teeth large, compressed, greatly elevated, being embedded in large sockets, which swell in a rugged manner from the upper surface of the rami, giving the mandible a peculiarly distinctive form.

---

[1] δίς, twice, ὅπλα, arms, and ὀδούς, tooth, that is, armed with two teeth.

DIOPLODON DENSIROSTRIS, de Blainville.   The Dioplodon.

Synonyms—*Ziphius densirostris*, de Blainville.

    Mesodiodon densirostris, Duvernoy.

    Dioplodon densirostris, Gervais.

    Dioplodon Sechellensis, Gray, S. & W., p. 355, Suppl. p. 102.

Teeth $\frac{0-0}{1-1}$, placed on the anterior third of the rami ; the posterior edge of the symphysis does not reach the teeth.

Inhab : Seychelles, Lord Howe's Island.

"The total length of the skeleton, without cartilage, is 14 feet 8 inches ; the head measures 2 feet 5¼ inches in length, and 14 inches across at the widest part ; the lower jaw 2 feet 3 inches long, and 6¼ inches high behind the tooth ; five anterior pairs of ribs are jointed to the sternum ; sternum composed of four pieces ; the left tooth measures 6 inches in length, 3⅜ in width, and 1¼ thick"; condensed from Mr. Krefft's account in the Pro. Zool. Soc., 1870.

This unique and valuable skeleton had been for many years lying in a neglected state on Lord Howe's Island, when it was seen by Mr. Edward Hill. The gentleman being aware of the rare nature of the remains, impressed upon the residents that, if the bones were carefully collected and taken to Sydney, he would guarantee to them a favourable sale, and advised them to apply first at the Museum, of which he is a Trustee. In consequence of this advice, the skeleton, nearly complete, reached Sydney, and Mr. Krefft at once secured it for the establishment over which he is the Curator.

Genus BERARDIUS, Duvernoy.

Dorsal fin small, subfalcate (?) ; skull, head small, upper portion nearly symmetrical ; anterior surface of the premaxillæ do not curve forwards ; beak subcylindrical , narrow, much elongated, nearly five-sevenths of the entire length of cranium ; mandibular symphysis moderate, about one-fourth of the entire length of the ramus, and not anchylosed ; teeth compressed, moderate sized, in front of the lower jaw only.

BERARDIUS ARNOUXI, Duvernoy.   New Zealand Berardius.

Synonyms—*Berardius Arnuxii*, Duvernoy, Gray, S. & W., p. 348 ; Suppl., p. 99. Hector, Knox, Haast, Trans. New Zealand Institute.

    Berardius Arnouxi, Flower, Trans. Zool. Soc., 1872.

Teeth $\frac{0-0}{2-2}$, smaller than in ziphius, frequently not passing through the gums.

" Colour deep velvet-black ; belly greyish ; pectoral fins a little above the middle of the body ; dorsal fin small, falcate."

"The animal has the power of protruding the four teeth at will; it was young; lived on cephalopods, for the stomach contained about a bushel of the horny beaks of the octopus, which were nearly all of the same size; and it measured in length 30 feet 6 inches.

"This whale was cast ashore on the coast of Canterbury, New Zealand." From Dr. Haast—1868.

Duvernoy's specimen, obtained in 1846 at the Port of Akaroa, was 32 feet long. A smaller one was stranded in 1862 on the west coast of New Zealand, and described by Dr. Hector and Mr. Knox; and in 1870 another animal of this species was captured near the entrance to Port Nicholson. This measured 27 feet, and is described in the Trans. of the New Zealand Institute, by Mr. Knox: "The tooth is still sheathed in the gum, being embedded in a tough cartilaginous sac, which adheres loosely in the socket of the jaw, and is moved by a series of muscular bundles that elevate and depress it."

There is no representative of this cetacean in the Australian Museum.

(c.) MICROZOOPHAGA[1] or Insect-eaters.

SUB-ORDER II. ANODONTOCETE.[2]

Whalebone Whales.

Mysticete, Gray; *Mystacoceti*, Flower; *Cete vermivora*, Lesson; *Cetacea edentula*, Brisson.

Teeth none; palate furnished with long whalebone, hollow at the growing end, functionally analogous to the pulp-cavity of the molar teeth of the megatherium, or the tusk of the elephant; whalebone disposed in numerous parallel laminæ, pendent in two longitudinal rows from the roof of the upper jaw; each blade composed of a central layer of course fibrous tissue, emitting from its inner edge fine hair-like filaments, and coated on both sides and outer edge by compact, more or less polished enamel; head exceedingly large; external respiratory organ divided into two distinct orifices; gullet very contracted; eyes small, near the angle of the mouth; upper jaw more or less arched on the roof, narrow, and commonly shorter than the lower one; lower jaw broad, greatly curved outwardly and receives the upper lips when the mouth is closed; rami of the mandible connected by fibrous tissue at their tips and not by a true symphysis; upper surface of skull symmetrical; sternum composed of one piece, and attached by bone direct to the first pair of ribs only, there being no costo-sternal ribs.

[1] μικρός, small; ζῶον, animal; and φάγω, I eat.
[2] ἀ, without; ὀδούς, tooth; and κῆτος, whale.

E. Teeth none, rudimentary and absorbed.

Family XIII. BALÆNOPTERIDÆ.[1]

Finner Whales.

Baleen[2] short, broad; dorsal fin distinct, compressed, falcate; pectoral fins comparatively moderate; fingers four; head moderate, elongate, flattened; body elongate; throat and belly deeply longitudinally plaited; skull broad, depressed.

Genus PHYSALUS,[3] Lacepede.

Dorsal fin high, erect; vertebræ, 61–64; ribs, fifteen or sixteen pairs; nasal bones short, broad, deeply hollowed on their superior surface and anterior border; rami of the lower jaw massive, with a very considerable curve, and a high pointed coronoid process; cervical vertebræ free; head of the first pair of ribs simple, articulating with a transverse process of the first dorsal vertebra; sternum broader than long, in the form of a short broad cross, resembling the heraldic trefoil, but subject to considerable individual modifications. (*Principally from Flower.*)

PHYSALUS[3] ANTIQUORUM,[4] Gray. The Razorback.

Synonyms—*Balæna physalus*, Linnæus.

> *Physalus antiquorum*, Gray, Pro. Zool. Soc., S. and W., p. 144; Suppl., p. 53, Flower, P.Z.S.
>
> *Physalus Duguidii*, Heddle.
>
> *Balænoptera musculus*, Flem., Cuv., Eschr., Lillj, Maling, Van Beneden, Gervais.
>
> *Pterobalænna communis*, Eschr., Van Beneden.
>
> *Razorback of the Whalers.*

Baleen, slate-coloured, under-edge blackish, inner-edge pale streaked; colour of the adult animal slate-grey, beneath whitish. "Where the body was black, the furrows and their interspaces were black also, being covered with skin of the same texture as the body. Where the black of the body began to wash off into the white of the lower parts, the furrows were black and the interspaces pure white. On the lower surface, where the colour was white, the plicæ when separated were lined with a rosy epidermis. The colour of the back of the head and of the sides, to a line passing from the tail beneath the pectoral, black. The jaws, and upper and under sides of both pectorals and tail, black. Scattered irregularly over the back were greyish spots, three or four in a square foot, resembling the appearance produced by touching the

---

[1] *Balæna*, and πτερόν, fin.
[2] From *Baleine*, or *fanou* (French), whalebone.
[3] From φυσάω, to blow or puff up.
[4] *Antiqui*, ancients. Probably this species was known to Aristotle and Pliny.

skin with a slightly whitened finger. The polished surface gave the whole body a greyish appearance, and it was said to be grey."[1]

The length of the Razorback usually varies from 60 to 80 feet, but in extreme age it will reach to over 100 feet.

Inhab. the North Sea, occasionally entering the Mediterranean.

Mr. Van Beneden considers that this species is probably identical with the whale described by Aristotle as a large cetacean, having within its mouth bristles like those of a hog, instead of teeth, and found occasionally in the waters of the Mediterranean. The surmise of this learned writer is corroborated in a measure by the fact that in modern times no other baleen-bearing whale is known to enter within this inland sea.

The common habitat of this species is in the higher latitudes of the North Sea, but many individuals during winter travel southward to more genial climes, and thus have been frequently captured on the coasts of Great Britain, France, and Holland, and sometimes within the Mediterranean.

The British Museum specimen is the skeleton of an animal said to be 102 feet in length, and which was found dead, floating on the sea in Plymouth Sound, in 1831. Another whale, about the same time and place, was discovered with its gullet filled with a large quantity of pilchards, by which it was supposed to have been choked.

In November, 1869, a fine example of this species was stranded at Longniddry, in the Frith of Forth, and a few days after the occurrence a very characteristic and picturesque representation of the animal, as it lay helpless on the beach, appeared in the *Illustrated London News*; but the delineation is not attended with that scrupulous accuracy of detail sufficient to meet the requirements of the matter-of-fact naturalist.

The length of this latter animal is recorded as being, in a straight line, 78 feet 9 inches, with a girt of 33 feet; the breadth of the forefin 11 feet, and that of the tail 15 feet. The colour is described as slate-grey, with whitish tints beneath. The lower jaw projected considerably beyond the upper one.

PHYSALUS SIBBALDII, Gray.   Sibbald's Finner.

Synonyms—*Physalus Sibbaldii*, Gray, S. & W., p. 160.

Physalus latirostris, Flower, P.Z.S., 1865.

Cuvierius Sibbaldii, Gray, S. & W., p. 380; Suppl., p. 54.

Balænoptera carolinæ, Malm.

Balænoptera Sibbaldii, Van Beneden & Gervais.

Balæna maximus borealis? Knox.

Great Northern Rorqual? Jardine, Nat. Libr. (Knox).

The Steypireyör of the Icelanders.

---

[1] "Orkney Whales"—Heddle, P.Z.S.

Baleen, black, short, and very broad at the base.
Inhab. North Sea ; ascending rivers.

The beak of the skull is of great breadth to half its length, whence it contracts towards the tip, not gradually tapering from the base, as in the preceding species. This peculiarity, and the wide cheek-bones, the sternum of an irregular oval, and two additional caudal vertebræ, form the distinguishing features between the present animal and the Razorback.

The colour of Sibbald's finner, the grey fin-whale of Turner, is of a deep brown, verging upon green ; in size it equals any known species of the sub-order. It is seldom taken by the whalers, because the inferior quality of the whalebone and small yield of oil are not commensurate with the risk of the capture.

Dr. Gray considers that the Great Northern Rorqual, figured in Jardine's Naturalist's Library, and so ably described by Dr. Knox, belongs to the above species, the Physalus (Cuvierus) Sibbaldii.

Of the Balænoptera Carolinæ, placed in the synonyms, Mr. Malm says that it bears on the skin the usual number of cirripeds, but within the body an intestinal worm was found of quite a new form, to which he has given the name of Echinorhynchus[1] brevicollis[2].

PHYSALUS PATACHONICUS, Burmeister.   The Buenos Ayres Finner.

Synonyms—*Balænoptera patachonica*, Burmeister, Van Beneden.

   Physalus patachonicus, Gray, S. & W., p. 374, Suppl., p. 53.

   Physalus australis, Gray, ? 1850.

   Balæna australis, ? Desmoulins.

Baleen black throughout.
Inhab. Southern and eastern coasts of South America.

This whale, of which a portion of the skeleton is only known, is distinguished from the Physalus antiquorum, by characters very similar to those exhibited by the Phys. Sibbaldii, namely, by the great breadth of the face of the skull, continued to half its length before it narrows to the tip of the muzzle, and by the lateral rings of the second, third, and fourth of the cervical vertebræ being shorter than the diameter of the body of the vertebræ.

Seeing that the sternum, an important part, is absent, it is difficult to say how this whale differs in its anatomy from the Phys. (Cuvierius) Sibbaldii.

However this may be, it appears to represent in the Southern Hemisphere, by its size and peculiar osseous structure, the Physalus Sibbaldii of the Arctic Seas.

---

[1] ἐχῖνος, hedgehog, and ῥύγκος, beak.
[2] *Brevis*, short, and *collum*, the neck. These worms have a proboscis armed with little bent hooks, by which they cling to the intestines, and frequently penetrate through them.

PHYSALUS ANTARCTICUS, Gray.   The Antarctic Finner.

Synonyms—*Physalus antarcticus*, Gray, S. & W., p. 164.
    Sibbaldius? antarcticus, Gray, S. & W., p. 381, Suppl., p. 55.
    Bolænoptera antarctica, Van Beneden.

"There has been imported from New Zealand a quantity of finner-fins, or baleen, which are all yellowish-white ; this doubtless indicates a distinct species." (Gray, p. 164.)
Inhab. Buenos Ayres.

Genus SIBBALDIUS, Gray.

Dorsal fin very small, far behind, and placed on a thick prominence ; vertebræ 56–58 ; ribs, fourteen pairs ; nasal bones elongate, narrow, flat, or very slightly hollowed on the sides of the upper surface ; lower jaw with a comparatively slight curve, and a low, obtusely pointed coronoid process ; cervical vertebræ free ; head of the first rib bifurcated, articulating with the seventh cervical and first dorsal vertebræ respectively ; sternum very small, short, broad, somewhat lozenge-shaped. (*Principally from Flower.*)

SIBBALDIUS LATICEPS[1], Gray.   Broad-headed Finner.

Synonyms—*Balæna rostrata*, Rudolphi.
    Rorqual du Nord, Cuvier.
    Pterobalæna boops, Eschricht.
    Sibbaldius laticeps, Gray, S. & W., p. 170.
    Balænoptera borealis, ou laticeps, Bened. & Gervais.
    Rudolphius laticeps, Gray, Suppl., p. 54.

The colour of this whale is black above and white underneath ; the pectoral fins, according to M. Van Beneden, are entirely black, and not relieved at their base by any white, as in the Pike-whale, B. rostrata.
The total lengths varied from 31½ to 40 feet, but the animals were young from whom the measurements were taken. The vertebræ, ribs, and other bones are small and of delicate structure.
Inhab. North Sea, Norwegian coasts, Zuyder Zee, &c.
This small cetacean was usually confounded with its smaller neighbour, the Balæna rostrata of Muller and O. Fabricius, until Dr. Gray, in 1846, detected and clearly pointed out the difference in their anatomy.

SIBBALDIUS BOREALIS[2], Lesson.   The Flat-back, or the Ostend Whale.

Synonyms—*Baleine d'ostend*, Van Breda.
    The Ostend Whale, Guide to Exhib. Char. Cross.
    Balænoptera gigas, Esch. & Reinh., 1857.
    Pterobalæna gigas, Van Beneden, 1861.
    Sibbaldius borealis, Gray, 1866, S. & W., p. 175. Suppl., p. 55.
    Flowerius gigas, Lilljeborg, 1867.

---

[1] *Latus*, broad, *ceps*, from *caput*, the head.
[2] *Borealis*, northern.

Inhab. North Sea.

Colour—black above, white underneath.

"A whale was observed floating dead in the North Sea between Belgium and England, and was towed into the harbour of Ostend on the 4th November, 1827. It was 102 feet long," or precisely of the same length as the Razorback. This huge animal is our Sibbaldius borealis, the skeleton of which was exhibited at Charing Cross, London, being previously well described and figured by MM. Dubar and Scharf respectively.

This specimen was a female whale, and had the upper jaw narrower and shorter than the lower, so as, when the mouth is shut, to be completely embraced within the latter. The dorsal fin was placed posteriorly at nearly three-fourths of the entire length of the body. The bifurcation, or rather double-head, of the anterior ribs, is well developed in the skeleton of this aged animal, corroborative of a distinct generic peculiarity, one only attributed by MM. Van Beneden and Gervais to the young, and that but occasionally.

Setting aside the characters displayed by the skeletons, I may remark, that in colour, habits, places of resort (with *perhaps* the single exception of the Mediterranean), and in rivalry for superiority of size, as to which may be looked upon as the most bulky and powerful of created beings, the Razorback and the Flatback bear so strong a resemblance as to render their non-identity a matter of great difficulty.

Sibbaldius Schlegelii, Flower. The Javan Finner Whale.

Synonyms—*Balænoptera sc. physalus*, Schlegel.
Balænoptera Schlegelii, Flower, Van Beneden.
Balænoptera longimana, Schlegel.
Sibbaldius Schlegelii, Gray, S. and W., p. 178, Suppl., p. 55, Flower, P.Z.S.

Inhab. coasts of Java.

Mr. Flower, in Proceedings of the Zoological Society of London for 1864, gives a lengthened account and many illustrations of the skeleton of this species, to which I must refer the student, as the limited space at my disposal will not permit me to extract so profusely as I could desire from so esteemed a writer.

This inhabitant of the tropical seas between the Indian and Pacific Oceans, is evidently the counterpart of the northern species, the Sibbaldius laticeps, with which indeed it agrees so closely in every constituent part of the skeleton, that Mr. Flower closes his most elaborate analysis with the observation: "In the present case I have carefully compared the skeletons (that from Java and those from the European coast) together. I have even had the advantage of placing many of the bones of the two in the Leyden Museum side by side; and I confess that, allowing for difference of age, it is difficult to fix upon any characters in which they decidedly differ."

Dr. Gray, however, points out that the beak of the skull in proportion to the length of the brain cavity is much longer in the Javan than in the Broadheaded Finner.

The Javan skeleton belongs to a somewhat larger animal than its European representative—one which would probably measure 45 feet in length.

Genus BALÆNOPTERA.[1] Beaked or Piked Whales.

Dorsal fin high, erect; vertebræ, 48–50; ribs, eleven pairs; nasal bones rather narrow and elongate, truncated at their anterior ends, convex on the upper surface in both directions; rami of lower jaw much curved and with a high coronoid process; cervical vertebræ partially anchylosed; head of the first rib simple; sternum longer than broad, having the form of an elongated cross. (*Principally from Flower.*)

BALÆNOPTERA[1] ROSTRATA,[2] Müller. The Pike Whale or Lesser Rorqual.

Synonyms—*Balæna rostrata*, Müller, O. Fabr, Hunter, Nilsson.
    Balænoptera acuto-rostrata, Lacep., Scoresby, Lesson.
    Rorqualus minor, Knox, Jardine, Nat. Libr.
    Balænoptera rostrata, Gray, S. and W., p. 188, Suppl., p. 56.
    Pterobalæna minor, et rostrata, Van Beneden.

Colour—black above, beneath reddish-white; pectoral fin white near the upper part of the base; length, from 25 to 30 feet.

Inhab. North Sea, ascending the mouths of rivers.

This small whale, the smallest among the Anodontocete, is very active in its movements, and well known to the coast inhabitants of the northern portions of Europe and America, for its range of habitat extends from the temperate parts of the Atlantic to beyond the icy waters of the Arctic Circle.

Like others of the family, this species exists principally on the smaller kinds of fish, such as the Arctic salmon,[3] the herring, &c., in the anxious pursuit of which it ascends the mouths of rivers, and occasionally gets entangled within the folds of the drift-nets set by fishermen to intercept the immense shoals of these gregarious fish.

In its habits the Lesser Rorqual may be considered as solitary, for rarely two or three are seen together; and M. Eschricht, who has paid great attention to the whales of the North Sea, records that its period of gestation is ten months.

---

[1] *Balæna*, and πτερόν, fin.

[2] *Rostrata*, beaked.

[3] The Arctic Salmon is the *Salmo Rossii*, and is readily known from the common salmon by the remarkable length of the lower jaw, which extends far beyond the upper one. It is " so extremely abundant in the sea, near the mouths of the rivers of Boothia Felix at certain seasons, that 3,378 were obtained at one haul of a small-sized sein. They varied in weight from 2 to 14 pounds, and rather exceeded in the aggregate 6 tons."

BALÆNOPTERA BONÆRENSIS, Burmeister. The Pike Whale of Kerguelen's Land.

Synonyms—*Balænoptera bonærensis*, Burmeister, P. Z. S., 1867. Ann. del Mus. Pub. de Buenos Ayres, 1868 : Van Beneden & Gervais, Ost. des Cét., 1870.

The Fin-backed Whale of Desolation, near *Kerguelen's Land.?* Nunn's Narrative.

Baleen short, narrow.

The colour black, but lighter underneath. Length from 30 to 32 feet.

The small fin-backed whales of Kerguelen's Land and the coast of Buenos Ayres, described respectively by Messrs. Nunn and Burmeister, appear to be identical, bearing a similar external form and dwarfed dimensions of the body, and moreover occupying nearly the same longitudinal belt of the South Atlantic.

The skeleton of the B. Bonærensis has been well defined by Burmeister and Van Beneden, and by their report it seems that the Pike Whale of Buenos Ayres is closely allied to the Pike Whale of the North Sea, the two differing but little in their organic structure. In both the beak is straight; the vertebral column is composed of forty-eight or forty-nine joints ; the second and third of the cervical vertebræ are partially anchylosed, the normal condition of the neck bones ; and the sternum exhibits the lengthened form of the Latin cross. These connecting characters are aided materially by the diminutive frame, and, though on opposite sides of the equator, by a similar geographic range of habitat.

Inhab. Southern Sea, coast of Buenos Ayres, Kerguelen's Land ?

BALÆNOPTERA SWINHOEI, Gray. The Chinese Finner.

Synonyms—*Balænoptera Swinhoei*, Gray, S. & W., p. 382 ; Van Beneden & Gervais, Ost. des Cét., 1870.

Swinhoia chinensis, Gray, Suppl., p. 57.

Inhab. Formosa.

" Mr. Swinhoe has sent to the British Museum part of the head, three cervical vertebræ, the first and seven other dorsal vertebræ, and eight ribs of a large Finner Whale, which was thrown ashore on the coast of Formosa. The bones are nearly of the size of similar bones of the European Finner (Physalus antiquorum), which often reaches to 60 or 70 feet, and they most probably belong to an animal nearly of that size."

" The second and third cervical vertebræ are united, as in the small Finner (Balænoptera rostrata) of Europe, while in all the larger Finners which are as yet known these two bones are always free. The union of the second and third cervical vertebræ is one of the characters by which the genus Balænoptera is separated from the genus Physalus."—*Gray*, p. 383.

This animal, although retained here as a species of Balænoptera, is now considered by Dr. Gray as the type of a new genus, the *Swinhoia*.

In closing this short history of the Finner whales, by far the most rapacious of the whalebone-bearing group, I may observe that the economy, with scarcely a single modification, of each individual, is characteristically portrayed in the following brief extract :—" In a glassy sea, near Wick, a Finner rushed round us in every direction, with its upper jaw above the water, blowing with great violence and noise, and diving sometimes tranquilly, sometimes in a seething wave created by its fins and tail. It was evidently feeding on herrings, as every now and then it would rush headlong into portions of the sea, where the smooth surface was broken by the shoals of fish. The blow-holes were at times flat and unprojecting, at others boldly prominent, the animal evidently having the power of raising or depressing these organs. The fin whales of Orkney and Caithness every season are observed in pursuit of herrings."—*Heddle*, P. Z. S., 1856.

Family XIV. MEGAPTERIDÆ.[1]

HUMP-BACKED WHALES.

Baleen short, broad, triangular, rather twisted when dry, edged internally with a series of rigid fibres ; dorsal fin, or rather hump, low, broad, placed behind the middle of the body ; pectoral fins, narrow, very long, nearly one-fifth of the entire length of the body; fingers four, very long; head broad, flattened, less than one-fourth the length of the body ; throat, chest, and part of the belly, deeply, broadly, longitudinally furrowed with dilatable folds of the skin ; body comparatively short and robust ; skull intermediate in form between the preceding and following families ; beak broad behind, contracted in front; lower jaw, slender, much arched, longer than the upper one ; cervical vertebræ commonly free.

MEGAPTERA BOOPS,[2] O. Fabricius. The Keporkak.

Synonyms—*Balæna boops*, O. Fabricius, Nilsson, Turton.
    Balæna longimana, Rudolphi.
    Megaptera longimana, Gray, S. & W., p. 119, Suppl., p. 50.
    Megaptera boops, Van Beneden and Gervais.
    The Keporkak of the Greenlanders.

Colour black, excepting the pectoral fins and belly, which are white, mottled, and streaked with black ; the lower lip is studded with two series of tubercles.

Length, from 45 to 60 feet.

Inhab. Coasts of Greenland, Norway, Baltic, Scotland, Bermudas, &c.

---

[1] μέγας, great, and πτερόν, fin.
[2] βοῦς ox, and ὄψ bellow, in allusion to the violent *blowing* of this species.

The food of this species consists of those small animals which infest in such vast assemblages the Northern Ocean, and among these the Mallotus[1] arcticus, Ammodytes[2] tobianus, and Limacina[3] arctica, are especially pointed out by Eschricht.

The Keporkak has the power, during ordinary or tempestuous weather, of turning a complete somersault in the air, a feat which it is said no other cetacean is able to perform.

MEGAPTERA LALANDII, Fischer. The Cape Humpback.

Synonyms—*Rorqual du cap*, Cuvier.
Rorqualus antarcticus, F. Cuvier.
Balæna Lalandii, Fischer.
Poescopia Lalandii, Gray, S. & W., p. 126, Suppl., p. 51.
Megaptera Novæ Zelandiæ, Gray, S. & W., p. 128, Suppl. p. 50.
Megaptera kuzira, Gray, S. & W., p. 130, Suppl., p. 50.

Baleen, colour bluish; laminæ, 300 on each side; length near the angle of the mouth, 1 foot.

The Cape humpback differs from the northern animal in the following particulars: the head is more depressed; the temporal bone broader; the tip of the lower jaw more acutely rounded; the cervical vertebræ more squarely moulded, with two or three of the anterior segments partially anchylosed; the pectoral fins longer; and the mandible, in proportion to the upper jaw, much longer and broader.

In other respects the two greatly resemble each other.

But another distinguishing feature might possibly be detected, when an opportunity occurs, in the form of the ear-bone, at present unknown, which may prove to be "shorter and more swollen" than that of the Keporkak, and, in fact, similar to the one possessed by the New Zealand species, of which it is the only known portion of the skeleton, (described and figured by Dr. Gray). On this account, and it appears a reasonable one, I place the New Zealand Humpback among the present synonyms.

Mr. A. Smith, who had an excellent view of an animal captured at the Cape, represents its external appearance thus: "Back and sides black; belly, dull white, with some irregular black spots; pectoral fins narrow, anterior and posterior edges irregularly notched, upper surface black, under surface pure white. Length from tip of lower jaw to hinder margin of tail-fin, 34½ feet."

---

[1] μαλλωτός, woolly, downy—so named from the fine teeth; a genus of the Salmonidæ, of which only one species, the *arcticus*, is known. This is a small fish, 6 to 7 inches long, with fine teeth, densely set as the pile in velvet; it is used largely as bait in the cod-fisheries.

[2] ἄμμος, sand, and δύτης, burrowing into. A genus of sand-eels, of which the *tobianus* is the *lesser sand-eel*, or *launce*. It is assumed that it was with the gall of this fish that Tobias anointed his father's eyes; hence the specific name.

[3] *Limacina*, a minute, marine, left-handed shell, of which only two species are known; both in their habits are gregarious and antipodean, and furnished with two comparatively large fins attached to the mouth.

M. Delande's specimen, the skeleton of which is in the Paris Museum, must have exceeded by several feet the preceding dimensions.
Inhab. South Atlantic, North, and South Pacific.

MEGAPTIRA AMERICANA, Gray. The Bermuda Humpback.
Synonyms—*Megaptera americana*, Gray, S. and W., p. 129. Suppl., p. 50.
  Megaptera burmeisteri, Gray, S. and W., p. 129. Suppl., p. 50.
  Physalus brasiliensis, Gray, S. and W., p. 162. Suppl., p. 53.
  Megaptera osphyia. Cope.
  The Norwega and The Mystica. Hartt.
Baleen, black, short, twisted.
Inhab. Western parts of the North and South Atlantic, Bermuda, &c.

In the western moiety of both Atlantics for a considerable latitudinal extent very large humpbacked whales are met with, whose external colouring and form, quality of baleen and osseous structure, are greatly alike. The group, however, has been partitioned by some eminent writers into several distinct species, as enumerated in the synonyms, but upon what grounds I cannot conceive, other than geographic range of habitat. For the investigations, severely critical, into their anatomy appear to have been directed more with the view of affording comparisons between them and the Keporkak and Cape Humpback, than with each other; consequently the references to the points of variation, detected in the skeletons, apply in their comparative estimates almost exclusively to the two species named. I can readily understand that these American Humpbacks may possibly differ from the Humpbacks of the Northern Seas and of the Cape of Good Hope, but, as yet, I know of no good reason why the M. Americana, M. Burmeisteri, M. Brasiliensis, and M. Osphyia should be generically or specifically parted.
These huge cetaceans derive their sustenance by preying upon the vast hordes of small beings of diversified natures congregated within and around the large area of gulf-weed[1] collected midway in the Atlantic

---

[1] The Gulf-weed Banks extend from 19° to 47° in the middle of the North Atlantic, covering a space almost seven times greater than the area of France. Columbus, who first met with the Sargassum about 100 miles west of the Azores, was apprehensive that his ships would run upon a shoal. The banks are supposed by Prof. Forbes to indicate an ancient coast-line of the Lusitanian land province, on which the weed originated. The late Dr. Harvey stated that species of Sargassum abound along the shores of tropical countries, the gulf-weed (Sargassum bacciferum) being found abundantly, occupying large spaces in various portions of the deep sea. This marine plant never produces fructification—the berries being air-vesicles, not fruit; yet it continues to grow and flourish, wholly propagated by breakage. Besides the one mentioned above, there exist considerable banks of the gulf-weed—at some distance from the Antilles—in the North Pacific, &c., and isolated specimens, carried up by the tide, have been found at the head of the Parramatta River, by the Rev. Dr. Woolls.

K

by the eddies of oceanic currents; by feeding upon the sea-wrack,[1] or may be upon the floating gulf-weed itself, or upon the lesser fish of gregarious habits, so profusely abundant along the whole line of coast.

The coral reefs which fringe the shores of Bermuda and Brazil seem to afford shelter to the female and her suckling.

Of the Megaptera Americana, Dr. Gray says: "This is doubtless the whale described in Phil. Trans., vol. 1, p. 11, 132, where an account is given of the method of taking it. It is described thus :—Length of adult 88 feet, the pectoral 26 feet (rather less than one-third of the entire length), and the tail 23 feet broad. There are great bends (plaits) underneath from nose to the navel; a fin on the back, paved with fat like the caul of a hog; sharp, like the ridge of a house, behind; head pretty bluff, full of bumps on both sides; back black, belly white, and dorsal fin behind," "they fed much upon grass (Zostera) growing at the bottom of the sea; in their great bag of maw he found two or three hogsheads of a greenish grassy matter." "Baleen from Bermuda, called Bermuda Finner, is extensively imported; it is similar to the baleen of the Grey Finner" (Cuvierius Sibbaldii, baleen of which is uniform deep black).

"The *Norwega*, M. Americana, is a humpback, which has a belly white and smooth (?), back very dark bluish; length, 50 to 55 feet. This whale gives more oil than the *Mystica*.

"The whalebone is short and sells well. The beach on which the whales are cut up is strewed during the season with bones. There must be the bones of 500 whales on the spot. The fishery is carried on at Bahia on a much larger scale than at Caravellas." "*Mystica* (M. Brasiliensis) differs from the *Norwega* in having the back black and the belly and throat furrowed. Sometimes there are white spots on the side.

"The first whales appear in the Abrolhos waters at about the end of May, and they stay until October.

"The females often bring young calves with them, and appear to seek the shelter of the reefs. The head-quarters of the Abrolhos fishery is at Caravellas, or, rather, at the mouth of the river Caravellas, where are situated the armaçoes or trying-houses.

"The fishery begins at Bahia, according to Castelnau, about the 13th of June, and lasts till the 21st September. At Caravellas, I was assured that the whales always appeared later than at Bahia, and the fishery does not always begin until the last week in June, continuing through the month of September."

It will be seen that the principal if not the only difference pointed out by Dr. Hartt between the *Norwega* and the *Mystica* in their living

---

[1] The various species of the sea-wrack are included in the Natural Order of Zosteraceæ, and seen at low water on the rocks of all countries of the world. In a recent state it is used largely as manure, its calcined ashes in the manufacture of glass, and in the early times of this Colony for stuffing beds, &c.

state, lies in the presence or absence of the longitudinal folds of the skin along the throat and belly, and this distinction having been founded upon a palpable error in respect to the first-named animal, utterly fails to prove their non-identity. As silence is generally understood to give consent, so we may conclude that the quality of the baleen, and the general habits of these cetaceans are the same ; for no allusion is made to either by this writer, so well acquainted with these whales, further than that the fisheries vary about a week in time at Bahia and Cara-vellas, a matter of but little consequence, except to show that the dates of the periodic migration to the Brazils coincide nicely with the time of the whaling season at Bermuda, viz., "from March to end of May, when they leave."—(Gray.)

Family XV. AGAPHELIDÆ.[1]

Scrag Whales.

Baleen narrow, short, and curved ; dorsal fin, none ; pectoral fin lanceolate, four-fingered ; head less than one-fourth of the entire length of body ; throat and chest smooth, not plaited ; cervical vertebræ, free ; body slender, elongate.

Agaphelus[1] gibbosus,[2] Erxleben.   The Scrag Whale.

Synonyms—*Balæna gibbosa*, Erxl.   Gray, S. & W., p. 90.

Agaphelus gibbosus, Cope.   Gray, Suppl. S. & W., p. 48.

Scrag Whale. Dudley.

"The baleen is peculiar ; throughout the length of the maxillary bone it nowhere exceeded 1 foot in length, and the width of the band, or length of the base of each plate, 4 inches.   It is of a creamy white ; the fringe very coarse, white, and resembling hog's bristles."   (Cope.)

"Instead of a fin upon its back, the ridge of the after-part of its back is scragged with half-a-dozen knobs or knuckles.   His bone (whalebone) is white and won't split."   (Dudley.)

Length, about 50 feet.

Inhab. North Atlantic.

Agaphelus glaucus, Cope.   Californian Grey Whale.

Synonyms—*Agaphelus glaucus*, Cope.

Rhachianectes glaucus, Cope.   Gray, Suppl., p. 48.

"145 laminæ of baleen on each side, the longest 18 inches long ; colour, bright yellow."   (Cope.)

Inhab. : California.

---

[1] ἀγα, greatly, and ἀφελής, simple ; *i.e.*, without folds on the throat and without a dorsal fin.

[2] From *gibbus*, a hunch or swelling on the back.

Family XVI. BALÆNIDÆ.

RIGHT WHALES.

Without dorsal fin or hump of fat; palate furnished with long baleen; head usually of more than a fourth of the entire length of body; belly and throat smooth, without longitudinal plaits or folds of skin; beak greatly arched, leaving a wide interval between the upper and lower jaws; pectoral fins short, truncated, very broad; fingers, five, short; cervical vertebræ anchylosed in one solid mass.

Genus BALÆNA,[1] Linnæus.

Baleen very long, thin, narrow at the base; elastic; enamel thick, polished; fringe long, fine, arranged in a single series; head about one-third of the entire length of the body; nasal bones long, attenuated in front.

BALÆNA[1] MYSTICETUS,[2] Linnæus. The Greenland or Right Whale.

Synonyms—*Balæna mysticetus*, Linn. O., Fabr., Lacep., Scoresby, Eschricht, Reinhardt, Lillej. Gray, S. and W., p. 81, Suppl., p. 38. Bened. and Gerv., Cét., p. 34.
    Balæna Greenlandica, Linnæus.
    Balæna angulata, Gray, Suppl., p. 39.
    Balæna nordcaper, Gray, Suppl., p. 39.
    The Right, or Whalebone Whale, Dudley, Scoresby.
    The Nord Kapper and *Nordcaper* of Egéde and Anderson.

The females of this species attain to a larger size than the males, exhibiting a condition of sexual growth the reverse of that shown by the sperm whale.

From the measurements of many specimens captured, the adult animal was found to vary from 50 to 65 feet in length. The upper portion of the head is high and narrow, but broadens greatly downwards, where, moulded by the mandible, it becomes broad and flattish : so much so, that, when viewed in front, the head presents a triangular form.

The bones of the skull are very porous, and thoroughly saturated with oil, and withal so very light as to float in water even when drained of the lighter material. The enormous head, from 15 to 20 feet long, 6 to 8 feet broad, and 10 to 12 feet high, presents when the mouth is opened a cavity as large as a room, and "capable of containing a ship's jolly-boat full of men." The plates of baleen, about 300 in each row, proceed from each side of the narrow upper jaw, and, spreading outwards, inclose at their lower ends the huge, soft, immovable tongue, presenting an ideal resemblance to the canvas falling from the tent-pole

---

[1] *Balæna*, Latin; φάλαινα, Greek, a kind of whale.
[2] μυστις, the nose of a large fish, and κῆτος, whale, in allusion to the very large head.

over a monster feather bed. The lower portions of the baleen are received within and protected by the lips of the mandible. The baleen itself originates from a thin, fleshy substance, resting upon the gum, and which affords a continuous supply of material requisite for its wonderful after-growth. In this species the whalebone reaches to from 9 to 12 feet in length. It is externally of a grey or greenish colour, while the fine fibrous filaments proceeding from its inner edge are black. These latter form a thick internal covering, which, acting as a screening apparatus, permits no particle to escape, but entangles and sifts the minute objects destined to be the support of this huge cetacean.

" The colour of the Greenland Whale is dark grey and white, with a tinge of yellow on the lower part of the head ; the back, upper part of the head, most of the belly, the fins, tail, and under part of the jaws, are deep black ; the fore part of the under jaw and a little of the belly are white, and the junction of the tail with the body, grey. They are sometimes piebald. Under-sized whales are almost entirely pale-bluish, and the suckers are of a pale blackish colour. The blubber is from 10 to 20 inches thick. The pectoral fins are from 4 to 5 feet broad, and 8 to 10 feet long. Tail, 20 to 30 feet wide."—Scoresby.

Inhab. the North Sea, between 65° and 78° latitude.

Incapable, from its toothless mouth and narrow throat, either of seizing a large prey, or of swallowing it, even when accidentally entrapped, the Right Whale is forced to live upon a group of very small, but, fortunately for itself, extremely abundant animals, whose entire lives are passed in the open seas, unsheltered save by the floating gulf-weed, and unprotected from the storm but by declining into the still waters of the deep.

These tiny creatures, a heterogeneous assemblage, mostly composed of peculiar shrimps, crabs, star-fish, and innumerable sea-snails, at the utmost of two or three inches in length, are but insignificant pigmies when compared with the bulk of the whale ; but they, small as they are, assume gigantic proportions as they feast upon those countless millions of microscopic beings, either individually invisible, or but imperfectly defined to our unaided eyes, which cover in the aggregate some 20,000 square miles of the surface of the open ocean, floating either in compact masses, or in those lengthened bands produced by oceanic currents, and forming the vast fields known from their colour to seamen as the *green waters* of the Arctic Seas.

In feeding, the lower jaw is let down and the rate of speed increased; the huge cavity thus urged along secures, like a fisherman's net, a rich harvest of insect game. This operation being often repeated, the combined proceeds of the several hauls serve at length to satisfy the capacious maw of the monster.

" The natural affection of this species is interesting. The cub, being insensible to danger, is easily harpooned, when the attachment of the mother is so manifested as to bring it almost certainly within the reach of the whaler. Hence, though the cub is of little value, it is often struck as a snare for the mother."

" There is something extremely painful in the destruction of a whale, when thus evincing a degree of affectionate regard for its offspring which would do honour to the superior intelligence of human beings; "yet," continues this otherwise humane writer, "the object of the adventure, the value of the prize, the joy of the capture, cannot be sacrificed to feelings of compassion."

Captain Scoresby, the able writer of the preceding paragraphs, and a practical and successful whaler, killed during twenty-eight voyages, no less than 498 whales, from whom he obtained 4,246 tons of oil and a large supply of whalebone; these together, realized a little over £150,000 sterling.

From 1814 to 1817, a period of great activity among whalers of all nations, the British alone captured in Greenland and Davis' Straits, 5,030 adults of this species, omitting of course the enumeration of the many helpless young, valueless, save as a lure for the destruction of their dams. This great and indiscriminate slaughter soon caused these localities, then crowded with these valuable animals, to be *fished out*, and the trade destroyed.

Genus EUBALÆNA,[1] Gray.

Baleen thick, moderately long, broad at the base; enamel thin; fringe arranged in several layers, coarse and rigid; head about one-fourth of the entire length of the body; nasal bones short and broad.

EUBALÆNA[1] BISCAYENSIS, Eschricht. The Bay of Biscay Whale.

Synonyms—*Baleine franche du golf de biscaye*, Eschricht.
   Baleine de biscaye, Van Beneden.
   Balæna biscayensis, Gray, S. & W., p. 89.
   Eubalæna biscayensis, Flower.
   Hunterius biscayensis, Gray, Suppl., p. 44.
   Balæna cisarctica, Cope.

This whale is of similar bulk to the Balæna mysticetus, but differs from it by the head being much smaller in proportion to the size of the body; by the baleen being shorter, more brittle, and thicker in substance; and by the habitat, that of the temperate regions of the North Atlantic, between the latitudes of 40° and 65°.

The skin also is said to be smoother, thicker, and of a bluish colour.

---

[1] εὖ, perfect, and *balæna*.

This whale, formerly abundant, is now rarely met with, the race having been in former times nearly exterminated by the Basque whalers; it is only known to modern science by the skeleton of a young animal, which was captured with its mother, at St. Sebastian, in 1834.

Inhab. Bay of Biscay.

EUBALÆNA JAPONICA, Lacepede.   The Japan Whale.

Balæna japonica, Lacepede, Gray, Beneden, and Gervais.

Balæna australis, Temminck.

Eubalæna Sieboldii, Gray, S. & W., p. 96; Suppl., p. 43.

The Japan whale, although formerly captured in numbers by the English, American, and Japanese whalers, is but very imperfectly known, for no reliable remains have been secured for the examination of competent men. Temminck and Lacepede give their account of it from a porcelain model, and drawings by Japanese artists; Eschricht and Reinhardt from an imperfectly developed fœtus, preserved in the Copenhagen Museum; and Dr. Gray, from specimens of the north-west coast whalebone in the British Museum. All these scientific men, however, agree in considering this animal as distinct from the Greenland and Cape whales, with whom it is commonly confused.

Temminck describes the general colour as black, with the belly and a spot over the eye, and another on the chin, white; Eschricht states, that he found the ribs to be greater in number than those of the B. Mysticetus; and Dr. Gray points out, in a clear manner, the distinguishing characters of the baleen.

"The baleen is nearly as long as the Greenland, varying from 7 to 12 feet long, and slender; but for the same length it is nearly twice as thick in substance, and it gradually diminishes in thickness towards the ends. The enamel, when the outer coat is removed, is not so polished as that of the Greenland, and when cut through, the central fibres are thicker, tubular, and occupy about one-fifth to one-eighth of the thickness—much more in proportion than they do in the Greenland fins, and the enamel and fibre are coarser in texture and much more brittle. The blades of this whalebone are generally flexuous, or *not kindly*, so that when cut into strips they have the defect of being variously bent and tapering towards the end, which, with their brittleness, greatly reduces their value."

Inhab. North Pacific Ocean, visiting periodically the Coasts of Japan.

The Japan whale rather excels the Cape whale in size, but in many respects bears a close resemblance to it.

EUBALÆNA AUSTRALIS, Desmoulins. The Cape Whale.

Synonyms—*Baleine du Cap*, Cuvier.
> *Balæna australis*, Desmoulins, Temminck, Gray, Beneden
> and Gervais.
> *Eubalæna australis*, Gray, S. & W., p. 91 ; Suppl., p. 43.
> *Hunterius Temminckii*, Gray, S. & W., p. 98; Suppl.,
> p. 44.
> *Caperea antipodarum*, Gray, S. &W., p. 101; Suppl., p. 45.
> *Balæna antipodarum*, Beneden & Gervais.
> *Neobalæna marginata*, Gray ; Suppl., p. 40.
> *Right whale of the Southern Seas*, Bennett, Polach,
> Crowther.

The Eubalæna australis (the Cape Whale), and the Caperea antipodarum (the New Zealand Whale) of Dr. Gray, are identical in size, in colour, in the quality of the baleen, and in the yield of oil. Both display similar habits, partake of the same kind of food, and, to the best of my belief, inhabit the same parallels of southern latitude. They are known to whalers, under the one name, as the Black or the Right Whale, of the Southern hemisphere. Nevertheless, the two are separated by Dr. Gray and M. Van Beneden into distinct species, from the following slight discrepancies in their structure, which, after all, may be "not greater than are found among different individuals of undoubtedly the same species."—Flower.

Dr. Gray, deprived of other osteological portions, bases his argument upon the form of an ear-bone, so distinctive as to necessitate even the creation of a new genus ; while M. M. Beneden & Gervais, more fortunate in possessing an entire skeleton for examination, wholly regard the *ear-bone* doctrine as visionary, and rest their claims for separation upon some slight variations in the proportions of the skull and in the number of vertebræ.

The French naturalists also lay great stress on the individual range of the species, which they most fancifully limit to two belts, embraced within the same parallels of latitude, and varying in breadth between five and six hundred miles, the Cape Whale occupying the one which stretches from the southern headland of Africa to the coasts of South America, and the New Zealand species being confined to the other, which extends from the west side of America to New Zealand ; the large intermediate space in the South Atlantic being, I presume, an interdicted locality. But as these whales, it matters not of which kind, notoriously frequented in former times the shores of Tasmania, in great numbers, and many were, and still are, captured even to the south of Kerguelen's Land, it will be at once seen by a glance at the map of the Southern Hemisphere that the latitudinal limit fixed upon by these authors has been thus more than doubled, and that their imaginary line of separation is simply incorrect.

The colour of the Right, or Black Whale of the Southern Seas, is, as the latter name implies, uniformly black; and in size it is somewhat inferior to the northern Right Whale.

In regard to their economy, for both so-called species are alike, I shall leave the several intelligent observers, residing widely apart, and well acquainted with their habits, to speak for themselves.

Mr. Warwick furnished Dr. Gray with the following observations and measurements of a female whale, taken at the False Bay Fishery, Cape of Good Hope:—

"Total length, 68 feet; width of tail, 15½ feet; diameter of gullet, 2 inches." "I could not pass my hand through the gullet." "These Whales of the Cape I constantly found covered with *tubicinella balænarum*[1] and *coronula balænaris;* but the Spermaceti Whale was seldom or never so covered; they occur principally on the head, where they are crowded." "They carry on the fishery from the shore, and only one bull out of sixty specimens was killed, the females coming into the bay to bring forth their young."

"The male whale (E. antipodarum)" says Dr. Dieffenbach, in his work on New Zealand, "is very rarely caught on the shores of New Zealand, as it never approaches the land so near as the females and young do. The season in which whaling is carried on is from May to October.

"In the beginning of May the females approach the shallow waters for the purpose of bringing forth their young. This period lasts about four months, as in May whales are seen with newly-born calves, and cows have been killed in July in full gestation." "The results of the whale fishery on the coast of New Zealand are of very small amount in the British market, owing to the indiscriminate slaughter of the fish during the last fifteen years, without due regard to the preservation of the dams and their young. The shore-whalers, in hunting the animal in the season when it visits the shallow waters of the coast to bring forth the young and suckle it in security, have felled the tree to obtain the fruit, and have thus taken the most certain means of destroying an otherwise profitable and important trade." "The beach at Tory Channel was covered with remains of whales' skulls, vertebræ, huge shoulder-blades and fins."

Dr. Crowther, of Hobart Town, Tasmania, whose science and zeal in matters connected with the Cetaceans of the southern seas are so

---

[1] The genera *tubicinella* and *coronula* are formed by small shell-like animals, which have their bodies and limbs articulated, and both protected by a conical, hard, external covering, also divided into segments. They belong to the class Cirrepeda of the sub-kingdom Articulata, but formerly were included by conchologists among the Mollusca. Several members of the group are found most abundantly adhering to rocks, timber, bottoms of ships, or on the backs of other living animals, attached either directly on their bases or by stems. The assertion in page 88 of the B. M. C., Seals and Whales, that each species of whale has its own peculiar kind of Sessile Cirrepede—one the *Coronula*, another the *Diadema*, and a third the *Tubicinella*—is not borne out by facts, as exemplified in the present, out of many instances that could be produced.

deservedly appreciated, has kindly supplied me with the following interesting particulars of the habits of this species :—

" This, the Eubalæna australis (Caperea antipodarum of Gray), is the Right Whale of the Southern Hemisphere. It is essentially a cold water fish, and has an almost unlimited geographical range, following the polar current almost to the edge of the tropics. It is also abundant in the South Atlantic. The American whalers, *en route* to the Pacific and Indian Oceans, take quantities of this oil at Tristan D'Acuna, the Crozets, and Kerguelen's Land.

" Some years since, these whales, for the purpose of calving, used to visit the bays and estuaries on the coast of Tasmania in great numbers, arriving in the month of May and leaving about the end of October. May, June, and the early part of July, are the calving months. A few males occasionally accompany them.

" So abundant were these fish that 100 were killed by the shore-parties at the Schouten Islands in four weeks. Each fish would yield 8 tons of oil and 7 cwt. of whalebone. In a season or two after this extraordinary slaughter, for the locality noted was only one out of several fisheries simultaneously in operation, this source of wealth and enterprise departed for ever from our shores."

ADDENDA.

THE recent arrival of the R.M.S. " China," bringing for our Museum two Parts of Volume VIII of the Transactions of the Zoological Society of London, enables me to give, as Addenda, more comprehensive and interesting accounts of the economy of the cetaceans Globiocephalus melas and Grampus griseus than the ones contained in pages 99 and 104 respectively.

I therefore extract the following information respecting these animals from the able writings of the eminent comparative anatomists, Mr. W. H. Flower, F.R.S., and Dr. James Murie, F.L.S.; although, in doing so, I regret being compelled to greatly condense the matter, and to alter the disposition of the text, in order to suit the nature of this treatise and the limited space to which I am restricted.

GLOBIOCEPHALUS MELAS, Traill.

For Synonyms, see page 99, to which add—

Globiocephalus melas, Murie. Trans. Zool. Soc., 1873, vol. 8, part 4.

Teeth $\frac{9\text{-}9}{6\text{-}9}$ to $\frac{12\text{-}12}{12\text{-}12}$.

" The numbers of teeth are most irregular, being so loosely implanted in their sockets that in early life, adolescence, and old age, they not unfrequently drop out."—*Murie.*

" Teeth in both jaws 9 to 12 on each side on the anterior half of the jaws, sometimes deciduous."—*Flower.*

" Colour deep black, partial whitish streak on the abdomen, narrowed posteriorly: some writers describe the colour as shining lustrous black, like *oiled silk.* When the surface of the skin is moist, it resembles the outer polish of fresh india-rubber; when dry, it becomes like lampblack, or of a sooty tint.

" The snout is very globose and prominent; the protuberant swelling projects nearly as far as the upper lip; the mandible, with its dense labial covering, is shorter than the upper lips; the dorsal fin is large, falcate, and laterally greatly compressed, and situated in front of the middle of the body; but the precise position of the cetacean dorsal fin would seem to be no sure specific test, for between fœtus and mother there is no unanimity; in other words, its position depends, *pari passu,* on the age of the animal; pectoral fins are low set, peculiarly narrow, tapering, and scythe-shaped.

"Above twenty of this species were killed in the Frith of Forth, in the latter end of April, 1867, the school containing from 150 to 200. The largest measured 26 feet long, and the smallest between 6 and 7 feet. Authentic accounts show that 40, 70, 92, 98, 150, and 200 have been destroyed at one onslaught."—*Selected from Murie*, T.Z.S., vol. 8, part 4.

GRAMPUS GRISEUS, Cuvier. Risso's Dolphin.

Synonyms—*Delphinus griseus*, Cuv.
      Delphinus aries, Risso.
      Grampus Cuvieri, Gray. S. & W., p. 295. Suppl., p. 83.
      Grampus Rissoanus, Gray. S. & W., p. 293. Suppl., p. 82.
      Grampus griseus, Flower. T.Z.S., vol. 8, part 1.

Teeth $\frac{0-0}{3-3}$ to $\frac{0-0}{7-7}$. "No traces of teeth in the upper jaw; 3 to 7 rather small teeth on each side of lower jaw, near the symphysis; apices worn down quite flat."

"The female, taken February, 1870, in a mackerel net, near the Eddystone Lighthouse, was young, but adult, and measured 10 feet 6 inches. Colour grey, varying in some parts to pure white, in others to deep black; but anteriorly the light parts had a yellowish wash, and the dark parts a slight bluish or purplish tinge. Anterior to the dorsal fin the colour is lightish grey, variegated with darker or whiter patches. On the top of the head there is a large, nearly black, patch, and the middle of the belly is greyish-white. The most remarkable characteristic was the presence of conspicuous, most irregular, light streaks and spots, scattered over the whole of the sides, from the front of the head to about 2 feet from the end of the tail. The streaks or lines were of various lengths, and running in the most fantastic manner, some parallel, some crossing each other, and some forming sharp angles, zigzags, and scribble-like patterns. These are entirely absent from the dorsal, pectoral, and upper surface of the caudal fins."

"This animal in its general form resembled Globiocephalus more than any other cetacean, having a similar rounded adipose protuberance, but developed to a less extent."

About a month after the capture of the above individual, another very young animal, also a female, was taken along the British coast, and exhibited in Billingsgate market, where it was subjected to Mr. Flower's examination. These two recently-acquired specimens agreed in their general colours, and in the peculiar disposition of the markings.

In order to institute comparisons between the foregoing animals and those previously described by Risso, Cuvier, Gray, and others, Mr. Flower gives a list of the various examples, stranded from 1822 to 1867, of Grampus griseus (said only to occupy the open seas), and of Grampus Rissoanus (hitherto only met with in the Mediterranean), for both have been considered as distinct species, from the variation in their colouring, and more particularly from the limited habitat assigned to each. These comparisons have perfectly justified Mr. Flower in uniting the two so-called species, and bestowing upon the animal a much wider range of habitat.

This able zoologist concludes his interesting remarks with :—" They (the two recently acquired), according to the coloration, should be *Rissoanus;* according to the habitat, they should be *griseus.* As to the teeth, the new specimens completely break down the specific distinction previously drawn; for, with the coloration of *G. Rissoanus*, the adult one has the number of teeth assigned to *G. griseus*, viz., $\frac{0-0}{3-3}$."

It thus appears necessary, until better diagnostic characters are made out, to sink the name of Rissoanus in that of griseus, though it may be convenient to apply the term Risso's Dolphin to the peculiarly marked variety which was first made known to science by that naturalist.—*Selected from Flower.* T.Z.S., vol. 8, part 1.

Habits migratory, visiting shores of Europe in summer, passing winter either to the south towards the coast of Africa, or to the west towards the American continent.

INDEX.

Sydney : Thomas Richards, Government Printer.—1873.